03-CTS-22UR

Fate of Pharmaceuticals and Personal Care Products Through Municipal Wastewater Treatment Processes

by:
Roger Stephenson, Ph.D., P.E., DEE
MWH

Joan Oppenheimer, M.S.P.H
MWH

2007

WERF

The Water Environment Research Foundation, a not-for-profit organization, funds and manages water quality research for its subscribers through a diverse public-private partnership between municipal utilities, corporations, academia, industry, and the federal government. WERF subscribers include municipal and regional water and wastewater utilities, industrial corporations, environmental engineering firms, and others that share a commitment to cost-effective water quality solutions. WERF is dedicated to advancing science and technology addressing water quality issues as they impact water resources, the atmosphere, the lands, and quality of life.

For more information, contact:
Water Environment Research Foundation
635 Slaters Lane, Suite 300
Alexandria, VA 22314-1177
Tel: (703) 684-2470
Fax: (703) 299-0742
www.werf.org
werf@werf.org

This report was co-published by the following organization. For non-subscriber sales information, contact:

IWA Publishing
Alliance House, 12 Caxton Street
London SW1H 0QS, United Kingdom
Tel: +44 (0) 20 7654 5500
Fax: +44 (0) 20 7654 5555
www.iwapublishing.com
publications@iwap.co.uk

Library of Congress Catalog Card Number: 2006928156
IWAP ISBN: 978-1-84339-772-4

MWH

The research on which this report is based was developed, in part, by the United States Environmental Protection Agency (EPA) through Cooperative Agreement No. CR-83155901-1 with the Water Environment Research Foundation (WERF). However, the views expressed in this document are solely those of MWH and neither EPA nor WERF endorses any products or commercial services mentioned in this publication. This report is a publication of WERF, not EPA. Funds awarded under the Cooperative Agreement cited above were not used for editorial services, reproduction, printing, or distribution.

ACKNOWLEDGMENTS

The authors are deeply grateful to the staff of the participating utilities for providing invaluable help throughout the different phases of this project, especially during sample collection. As some of these facilities agreed to collaborate under the condition of anonymity, their names are not included in this report. Our gratitude is also extended to the members of the project Technical Advisory Committee, Dr. Shane Snyder (Southern Nevada Water Authority), and Dr. R. Rhodes Trussell (Trussell Technologies Inc) for their helpful guidance and advice.

Report Preparation

Principal Investigators:
Joan Oppenheimer, M.S.P.H.
Roger Stephenson, Ph.D., P.E., DEE
MWH

Project Team:
Arturo Burbano, Ph.D.
Li Liu, M.Sc.
Suvid Surendran, M.Sc.
MWH

Greg Loraine, Ph.D.
Dynaflow

Project Subcommittee

Kenneth E. Banks, Ph.D.
City of Denton, TX

Rolf Halden, Ph.D., P.E.
Johns Hopkins University

John T. Novak, Ph.D., P.E.
Virginia Tech

Neil J. Parke
Eli Lilly & Company

Bernard M. Sawyer
Metropolitan Water Reclamation District of Greater Chicago

David L. Sedlak, Ph.D.
University of California, Berkeley

Water Environment Research Foundation Staff

Director of Research: Daniel M. Woltering, Ph.D.
Program Director: Christy Terhune, P.E.

ABSTRACT AND BENEFITS

Abstract:

Municipal wastewater treatment facilities were not specifically designed to remove compounds such as endocrine disrupting compounds (EDCs) and pharmaceuticals and personal care products (PPCPs). Previous studies predominantly in Europe have described the need for a minimum critical solids retention time (SRT) to achieve good removal of EDCs and pharmaceuticals. These studies have presented limited data on PPCPs from full-scale facilities. The objective of this study was to expand published findings to the removal of 20 PPCPs commonly found in the influent of full-scale treatment facilities operating in the United States. The six selected facilities cover operating SRT conditions from 0.5 to 30 days and include facility capacities in excess of 300 mgd.

The results demonstrate a wide distribution in both the occurrence of these compounds in the influent to secondary treatment processes and the percentage removal achieved through secondary treatment. The 20 PPCPs were categorized into a 3x 3 matrix of nine bin combinations of occurrence and treatment reduction. Although a large number of these compounds were well removed, one compound in particular (Galaxolide) occurred frequently and was resistant to removal. A minimum critical SRT, defined in this study as the minimum time needed to consistently demonstrate greater than 80% removal ($SRT_{80\%}$), was determined for the 20 target compounds. $SRT_{80\%}$ was compound dependent, with most compounds consistently removed in systems with a SRT of 5 to 15 days.

Benefits:

- Provides aqueous phase secondary process influent occurrence and treatment removal data for PPCP compounds commonly used in the United States.
- Demonstrates the benefit of the longer SRT used for nitrogen removal in reducing aqueous phase PPCP compounds before effluent discharge.
- Informs the wastewater treatment industry about specific compounds that are more difficult to remove through a secondary treatment process and their frequency of occurrence.
- Suggests the possibility of seasonal effects on SRT requirements for PPCP removal.

Keywords: secondary treatment, solids retention time, full-scale treatment plants.

TABLE OF CONTENTS

Acknowledgments iii
Abstract and Benefits iv
List of Tables vii
List of Figures viii
List of Acronyms ix
Executive Summary ES-1

1.0 Introduction 1-1

2.0 Materials and Methods 2-1
- 2.1 Project Approach 2-1
  - 2.1.1 Target Compounds 2-1
  - 2.1.2 Participating Facilities 2-2
  - 2.1.3 Membrane Bioreactor Pilot Systems 2-6
- 2.2 Sampling Protocols 2-7
- 2.3 Analytical Protocols 2-9

3.0 Results and Discussion 3-1
- 3.1 Occurrence of Soluble PPCPs in Sampled WWTPs 3-1
- 3.2 Removal of Soluble PPCPs by Secondary Wastewater Treatment 3-6
  - 3.2.1 Plant Operation and Performance Data 3-6
  - 3.2.2 Removal of Target PPCPs through Secondary Treatment 3-7
  - 3.2.3 Impact of SRT on the Removal of Soluble PPCPs 3-10
  - 3.2.4 Conventional Activated Sludge Processes versus Membrane Bioreactor 3-16
  - 3.2.5 Effect of Media and Membrane Filtration 3-16
- 3.3 Qualitative Observations 3-16
  - 3.3.1 Seasonal Variations of PPCP Occurrence 3-17
  - 3.3.2 Impact of HRT on the Removal of Soluble PPCPs 3-18
  - 3.3.3 Other Factors Potentially Impacting the Removal of PPCPs 3-20

4.0 Conclusions and Recommendations 4-1
- 4.1 Occurrence 4-1
- 4.2 Treatment 4-2
  - 4.2.1 Dependence of PPCP removal on SRT 4-3
  - 4.2.2 Other Conclusions 4-4
- 4.3 Future Work 4-5

Appendix A Sampling Equipment Specs A-1
Appendix B Sampling Protocol for 24-Hour Time-Composite Samples B-1
Appendix C Operational Data Log Sheets for Plant Personnel C-1
Appendix D Operational Parameters of Secondary Treatment and Water Quality Analysis of Secondary Influent/Effluent D-1
Appendix E General QA/QC Procedures E-1
Appendix F Method Detection Limit Study F-1

Appendix G Percent Removal of PPCPs through Tertiary Filters G-1
Appendix H Occurrence of PPCPs in Influent to Plants without Primary Treatment H-1
Appendix I Occurrence of PPCPs in Influent to Plants with Primary Treatment I-1

References R-1

LIST OF TABLES

2-1 Selected Target Analytes 2-2
2-2 Descriptions of Participating Treatment Facilities 2-3
2-3 Sampling Locations at Participating Treatment Facilities 2-5
2-4 Retention Times, Major Ions for Each Analyte, and SIR Segment Used for Quantification 2-10
2-5 IDLs and Calibration Curve Root Mean Squares for Each Analyte 2-11
2-6 Method Detection Limits and Ratios of the MDL to the Initial Spiked Concentrations for Each Analyte 2-12
3-1 Definition of Occurrence Bins 3-1
3-2 Concentration of PPCPs at Different Occurrence Percentiles 3-6
3-3 Definition of Treatment Bins 3-7
3-4 Percent Removal and Treatment Bins of Target PPCP Compounds 3-7
3-5 Distribution of Target PPCPs in Occurrence and Treatment Bins 3-9
3-6 Critical SRTs to Effectively Remove Target PPCP Compounds 3-15
3-7 K_{ow} of PPCP Compounds Classified in Different Treatment Bins 3-20
4-1 Classification of Analyzed PPCPs in Occurrence Bins 4-1
4-2 Classification of Analyzed PPCPs in Treatment Bins 4-2
4-3 PPCPs of Concern Based on Occurrence and Treatment Bins 4-3

LIST OF FIGURES

2-1 Schematic of Plant A (SRT ~0.5–1.5 days) 2-4
2-2 Schematic of Plant B (SRT ~3–5 days) 2-4
2-3 Schematic of Plant C (SRT ~5–6 days) 2-4
2-4 Schematic of Plant D (SRT ~7–20 days) 2-4
2-5 Schematic of Plant E (SRT ~11–16 days) 2-5
2-6 Schematic of Plant F (SRT ~25–30 days) 2-5
2-7 Schematic of the MBR Pilot #01 2-6
2-8 Schematic of the MBR Pilot #02 2-7
3-1 Cumulative Probability Distribution for the Occurrence of Bin O1 (Infrequent Occurrence) Compounds in Influent to Secondary Treatment 3-2
3-2 Cumulative Probability Distribution for the Occurrence of Bin O2 (Intermediate Occurrence) Compounds in Influent to Secondary Treatment 3-3
3-3 Cumulative Probability Distribution for the Occurrence of Bin O3 (Frequent Occurrence) Compounds in Influent to Secondary Treatment 3-4
3-4 Percent Removal in Relation to SRT for Compounds Susceptible to Activated Sludge Treatment (Bin T1, Good Removal) 3-11
3-5 Percent Removal in Relation to SRT for Compounds Susceptible to Activated Sludge Treatment (Bin T2, Moderate Removal) 3-13
3-6 Percent Removal in Relation to SRT for Compounds Susceptible to Activated Sludge Treatment (Bin T3, Poor Removal) 3-14
3-7 Occurrence of 3-Phenylpropionate in Plants without Primary Treatment. 3-17
3-8 Occurrence of 3-Phenylpropionate in Plants with Primary Treatment. 3-18
3-9 Correlation of HRT and SRT in Sampled Facilities 3-19
3-10 Percent Removal in Relation to HRT for Galaxolide and Musk Ketone 3-19
3-11 Seasonal Variation in the Removal of Musk Ketone, Galaxolide, and TCEP in Plant F 3-21

LIST OF ACRONYMS

BHA	butylated hydroxyanisole
BOD	biochemical oxygen demand
CAS	Chemical Abstracts Service
DEET	*N,N*-diethyl-3-methylbenzamide
EDC	endocrine disrupting compound
GC	gas chromatography
HRT	hydraulic retention time
K_{ow}	octanol-water partition coefficient
IDL	instrument detection limit
MBR	membrane bioreactor
MDL	method detection limit
MLE	modified Ludzack Ettinger process
MLSS	mixed liquor suspended solids
MS	mass spectrometry
NH_3–N	ammonia (as nitrogen)
NO_3–N	nitrate (as nitrogen)
NO_2–N	nitrite (as nitrogen)
PPCP	pharmaceutical and personal care product
RO	reverse osmosis
SIM	selective ion monitoring
SIR	select ion reporting
SPE	solid phase extraction
SRT	solids retention time
$SRT_{80\%}$	critical SRT (SRT required for achieving minimum removal of 80%)
TCEP	tri(chloroethyl) phosphate
TIC	total ion count
TKN	total Kjeldahl nitrogen
TSS	total suspended solids
U.S. EPA	United States Environmental Protection Agency
VSS	volatile suspended solids
WWTP	wastewater treatment plant

Executive Summary

ES.1 Background

In the past decade, a great deal of attention has been focused on the presence of a particular group of compounds in wastewater. These compounds include the endocrine disrupting compounds (EDCs) and pharmaceuticals and personal care products (PPCPs). EDCs are compounds that can interfere with natural hormonal reactions in living organisms by supplementing, mimicking, or antagonizing the hormones orchestrating these reactions. PPCPs are a broad class of compounds marketed to improve human health or used as key components of cosmetics, toiletries, and household consumer products. Some of these compounds require extensive testing before their registration and use. Although the human health impacts of many of these compounds are understood within the context of their intended use, their environmental impacts to wildlife and the multigenerational human impacts from widespread environmental exposure requires more study.

Secondary municipal treatment was developed to reduce the organic load of discharged effluent to minimize loss of oxygen in receiving waters. The biological activated sludge process is the predominant form of secondary treatment used throughout the world. There is a variety of activated sludge processes, ranging from short duration, high rate processes to biological nutrient removal and extended aeration activated sludge with long detention times. Although much is known about these processes, their ability to remove EDCs and PPCPs is not well understood.

ES.2 Project Approach

The primary objective of this study was to evaluate, through collection and analysis of samples from a range of full-scale activated sludge treatment facilities, the fate of aqueous phase PPCPs as a function of process solids retention time (SRT). The study also included evaluation of PPCP removal through two different pilot-scale membrane bioreactors (MBRs). The participating facilities and MBR units (presented in Table ES-1) cover a 0.5–30 day range of SRT operating conditions, and the collected samples were analyzed for a target list of 20 PPCP compounds (Table ES-2) using gas chromatography (GC) and mass spectrometry (MS) with selective ion monitoring (SIM; GC-MS-SIM). The samples were collected as 24-h time-weighted composites using refrigerated samplers equipped with stainless steel and Teflon tubing.

ES.3 Sorting PPCP Compounds by Occurrence and Removal Behavior

The aggregate data analysis sorted the target compounds into bin classifications on the basis of their frequency of occurrence in the secondary process influent and their degree of removal through the secondary treatment process. The three occurrence bins are defined in accordance with Table ES-3 and the three removal bins are defined in accordance with Table ES-4. The target compounds distribute amongst the nine different bin combinations of occurrence frequency and treatment removal as shown in Table ES-5. Compounds of greatest concern fall in the bin classification of frequent occurrence and poor removal. Galaxolide, a fragrance used in a wide variety of consumer products such as musk perfumes and laundry detergents, is the only compound to fall into this classification category

Table ES-1. Descriptions of Participating Treatment Facilities.

Facility	Primary Treatment	Secondary Treatment	Secondary Aeration	MLSS[1] (mg/L)	SRT (days)	Filters	Disinfection
A	Polymer ferric	High- purity O_2-activated sludge	Pure O_2	1,300–2,600	0.5–1.5	None	None
B	No chemicals	MLE[2] Nit./Denit.	Diffused air	1,800–2,000	3–5	Deep bed[3]	Chlorine
C	No chemicals	Activated sludge	Diffused air	2,000–3,000	4–6	Deep bed[3]	UV
D	No chemicals	Nit./Denit.	Diffused air	2,500–3,000	7–20	Granular microfiltration /RO	Chlorine
E	None	Nit./Denit.	Diffused air	2,100	11–16	None	UV
F	None	Ext. aeration Nit./Denit.	Surface air	4,000	20–30	Deep bed[3]	UV
MBR #01[4,5]	n/a	Nit./Denit.	n/a	14,000	14	n/a	n/a
MBR #02[6,7]	n/a	n/a	n/a	11,500	15	n/a	n/a

[1] MLSS, mixed liquor suspended solids; [2] MLE, modified Ludzack Ettinger process; [3] granular media; [4] Uses flat sheet membranes; [5] Located at plant E; [6] Uses free-end hollow fiber;
[7] Located at a facility not listed above.

Table ES-2. Selected Target Analytes.

Compound	CAS Number	Classification	MDL (µg/L)[1]	Log K_{ow}[2]
Galaxolide	1222-05-5	Fragrance	0.61	5.90
Ethyl 3-phenylpropionate	2021-28-5	Fragrance	0.012	2.73
Methyl 3-phenylpropionate	103-25-3	Fragrance	0.043	2.32
Musk ketone	81-14-1	Fragrance	0.034	4.30
3-Phenylpropionate	501-52-0	Fragrance	11	1.80
Benzophenone	119-61-9	Sunscreen	0.043	3.18
Octyl methoxycinnamate	5466-77-3	Sunscreen	0.11	2.30
Oxybenzone	131-57-7	Sunscreen	0.060	3.79
Benzyl salicylate	118-58-1	Sunscreen	0.022	4.31
Ibuprofen	15687-27-1	Pharmaceutical	1.4	1.07
Caffeine	58-08-2	Pharmaceutical[3]	0.053	-0.07
Triclosan	3380-34-5	Germicide	0.35	4.76
Chloroxylenol	88-04-0	Germicide	0.015	3.27
Tri(chloroethyl) phosphate (TCEP)	115-96-8	Fire retardant	0.18	1.44
Triphenylphosphate	115-86-6	Fire retardant	0.62	4.59
Methylparaben	99-76-3	Antioxidant	0.30	1.96
Butylated hydroxyanisol (BHA)	25013-16-5	Antioxidant	0.011	3.50
Butylbenzylphthalate	85-68-7	Plasticizer	0.13	4.91
N,N-Diethyl-3-methylbenzamide (DEET)	134-62-3	Insect Repellent	0.086	2.18
Octylphenol[4]	27193-28-8	Surfactant (detergent)	0.0040	4.12

[1] See Table 2-6; [2] Footnote references in Table 3-7; [3] Also from nonpharmaceutical sources; [4] n-octylphenol if not specified.

Table ES-3. Definition of Occurrence Bins.

Bin Number	Description	Bin Assignment Criteria
O1	Infrequent occurrence	Detected in less than 25% of the observations in influent
O2	Intermediate occurrence	Detected between 25% and 75% of the observations in influent
O3	Frequent occurrence	Detected in more than 75% of the observations in influent

Table ES-4. Definition of Treatment Bins.

Bin Number	Description	Bin Assignment Criteria
T1	Good removal	Median of the percent removal values is greater than 80%.
T2	Moderate removal	Median of the percent removal values is between 50% and 80%
T3	Poor removal	Median of the percent removal values is less than 50%

Table ES-5. Distribution of Target PPCPs in Occurrence and Treatment Bins.

Treatment / Occurrence	Bin T1 Good Removal	Bin T2 Moderate Removal	Bin T3 Poor Removal
Bin O1[1] Infrequent	Methyl-3-phenylpropionate	Octylphenol	TCEP Triphenylphosphate
Bin O2 Intermediate		Ethyl-3-phenylpropionate	**BHA** **DEET** **Musk ketone**
Bin O3 Frequent	Caffeine Ibuprofen Oxybenzone Chloroxylenol methylparaben Benzyl salicylate 3-Phenylpropionate butylbenzyl phthalate Octylmethoxycinnamate	**Triclosan** **Benzophenone**	**Galaxolide**

[1] For compounds in Bin O1 (infrequent occurrence), treatment classification is limited by insufficient data as they were seldom detected in the influent.

ES.4 Relationship of Treatment SRT and Removal of PPCP Compounds

Compound removals within each of the three treatment bin classifications were further analyzed to assess their dependence on secondary treatment SRT values. A critical SRT value was defined for each compound as the minimum SRT at which greater than 80% removal was consistently observed ($SRT_{80\%}$). A tabular summary of $SRT_{80\%}$ values observed for the target compounds is provided in Table ES-6. Furthermore, the collected data demonstrate no removal advantage for a pilot-scale MBR operating at an equivalent SRT value as a full-scale conventional activated sludge facility. This decline or increase was frequently, but not always, greater than the variability observed for plants sampled within the same month. Although it is possible to speculate on causes for this observation, a greater database should first be obtained...edecline

There was no relationship apparent between the octanol-water partition coefficients (K_{ow}) of the target compounds and their treatment removal bin classification. This was expected, because the study only included analysis of the aqueous phase of the treatment process.

ES.6 Observations Related to Treatment Hydraulic Retention Time (HRT) and Filtration

Several plants were operated at similar SRT values and dissimilar HRT values. This occurred because of the inclusion of two pilot MBR systems and one full-scale plant operating below design capacity. The limited data demonstrated no impact of varying HRT on the removal of the PPCP compounds.

Media filtration results showed a great deal of variability, but the overall data set demonstrated no evidence of PPCP compound removal caused by this process. Reverse osmosis (RO) treatment, however, was very effective in removing these same compounds.

Table ES-6. Critical SRTs to Effectively Remove Target PPCP Compounds.

Bins	Compounds	$SRT_{80\%}$ (days)
Bin T1 Good removal	Chloroxylenol	5
	Butylbenzyl phthalate	5
	Oxybenzone	5
	Octylmethoxycinnamate	5
	Benzyl salicylate	5
	Caffeine	5
	Ibuprofen	5
	3-Phenylpropionate	<5
	Methylparaben	<5
	Methyl-3-phenylpropionate	<5
Bin T2 Moderate removal	Benzophenone	13
	Triclosan	10
	Ethyl-3-phenylpropionate	>5
	Octylphenol	5~28
Bin T3 Poor removal	Musk ketone	>30[1]
	Galaxolide	>30[1]
	TCEP	>30[1]
	DEET	>13
	BHA	>8
	Triphenylphosphate	>5

[1] Although $SRT_{80\%}$ value is indeterminate, consistently good removal was not observed even in the plant operated at SRT as high as 30 days.

ES.7 Observations Related to Treatment Performance and SRT Values Below 5 Days

With the exception of three compounds (3-phenylpropionate, methylparaben, and methyl-3-phenylpropionate), an SRT ≥ 5 days was required to achieve consistent compound removal above 80% (i.e., $SRT_{80\%}$ = 5 days) for the compounds studied. For the two facilities operating below 5 days, much better removal was observed for the facility using high-purity oxygen.

ES.8 Key Study Conclusions

- In general, it was observed that an increase in SRT enhanced the removal of the majority of monitored PPCPs
- The SRT required to achieve consistent removal above 80% ($SRT_{80\%}$) is compound-specific with many of the target compounds well removed by activated sludge processes with SRTs of 5–15 days.
- An $SRT_{80\%}$ of more than 30 days was observed for Galaxolide, musk ketone, and tri(chloroethyl) phosphate.
- Activated sludge treatment can be effective in removing many PPCPs. However, to achieve good removal of most target compounds, a second treatment barrier might be necessary.

CHAPTER 1.0

INTRODUCTION

Municipal wastewater treatment facilities in the United States must comply with discharge limits for biochemical oxygen demand (BOD), total suspended solids (TSS), and other conventional pollutants. The core treatment used at these facilities is biological secondary treatment, with most of the largest metropolitan facilities using an activated sludge process. Supplemental or enhanced treatment is practiced at facilities subject to more stringent discharge requirements or those that produce effluent for reuse applications. Some new compounds may become of concern to regulatory agencies in the future, based on the availability of sufficient evidence (i.e., occurrence and toxicological studies) to justify their inclusion (Halling-Sørenson et al., 1998).

Among the compounds that may become regulated in the future, natural and synthetic chemicals collectively known as endocrine disrupting compounds (EDCs) and pharmaceutical and personal care products (PPCPs) are potential candidates. Studies supporting the theory that some of these chemicals can mimic the activity of natural hormones have existed for more than 70 years, and target these compounds as suspected causative agents of reported episodes of disruption in wildlife reproductive health (Snyder et al., 2003a). Although contamination from these chemicals may originate from non-point sources, a significant fraction comes from municipal wastewater treatment plant (WWTP) discharges (Daughton & Ternes, 1999). Municipal WWTPs can act as persistent point sources of EDCs and PPCPs, and trace concentrations of these chemicals have been observed in conventional secondary and tertiary wastewater discharges in the United States and abroad (Clara et al., 2005; Joss et al., 2004; Eriksson et al., 2003; Snyder et al., 2001).

EDCs and PPCPs have received less scrutiny in the past, mostly because of analytical limitations that prevented accurate detection and quantification of trace concentrations of these compounds (Ollers et al., 2001; Osemwengie & Steinberg, 2001). The advancement of analytical techniques such as gas chromatography (GC) and liquid chromatography in tandem with mass spectroscopy (MS) allows environmental scientists to identify and quantify these compounds at parts per billion (ppb or μg/L) or parts per trillion (ppt or ng/L) (Sedlak et al., 2000). The availability of more effective analytical tools has facilitated extensive research into the fate of EDCs and PPCPs in the environment, as these chemicals may pose a complex concern for two main reasons: (1) their effects are likely to occur at trace concentrations, and (2) their presence in effluent from municipal WWTPs is mostly caused by activities of individuals rather than regulated industrial discharges (Snyder et al., 2001; Daughton & Ternes, 1999). Consequently, understanding the ability of conventional WWTPs in preventing the passage of these compounds into the environment has become a critical concern (Clara et al., 2004).

Extensive research has been conducted to evaluate the occurrence and fate of hormonal EDCs in the environment. Estrogenic hormones (such as estradiol, estrone, and ethinyl estradiol) in particular, have been the target of multiple studies, as they have been suspected of being related to ubiquitous episodes of intersex in male fish in multiple locations in the United States and Europe (Snyder et al., 2004; Ternes et al., 1999; Irwin, 1998). Synthetic PPCPs remain a category considerably less studied, and comprehensive information on manufacturing, utilization, and disposal of these chemicals into the environment is less readily available. The findings of a United States Geological Survey reconnaissance survey of U.S. urban area streams (Kolpin et al., 2002) and other studies (Eriksson et al., 2003) have recently demonstrated the widespread environmental occurrence of PPCP compounds. These findings have generated concern by many parties about the presence of trace concentrations of these chemicals in the environment (Wilson et al., 2003). In addition, a subset of these PPCPs ultimately may be classified as EDCs.

This study surveyed a target list of PPCP compounds (i.e., covering a range of chemical characteristics) that have routinely been detected in wastewater influent, and assessed their removal through conventional secondary treatment facilities, as these processes constitute the cornerstone of municipal wastewater treatment systems. The data were then evaluated to determine if the removal of these compounds is influenced by the solids retention time (SRT), which is the master variable in the operation of activated sludge secondary treatment. In addition, PPCP removal through subsequent tertiary media filtration and disinfection, and newer hybrid treatment processes such as membrane bioreactors (MBRs) were also evaluated.

It has been reported that increasing SRT has a beneficial effect on the removal of PPCP compounds (Clara et al., 2005; Joss et al., 2004; Kreuzinger et al., 2004; Kanda et al., 2003; Ternes et al., 1999). Extended SRTs promote the growth of a more diverse biological community that is able to degrade PPCP compounds more efficiently because of co-metabolic effects. These results also suggest the existence of a critical SRT value after which, the removal of these compounds is not improved (Clara et al., 2005; Joss et al., 2004). These conclusions have been derived from studies conducted at small-scale WWTPs and bench/pilot studies performed under controlled conditions. The majority of these studies have been conducted in Europe, and have been more focused on estrogens and prescription pharmaceuticals than personal care products (Kreuzinger et al., 2004, Andersen et al., 2003; Ternes et al., 1999). For this reason, one of the main objectives of this research was to validate these conclusions using data from a group of large-scale WWTPs located in the United States, which includes facilities with capacities as high as 350 mgd. This study provides a systematic analysis of the effect of SRT on PPCP removal, by analyzing data from six full-scale WWTPs and two pilot-scale MBRs operating over a wide range of SRT values. Development of this information may prove valuable in defining optimum operational parameters for existing WWTPs, and the potential design of future installations, from the perspective of PPCP removal. This study will also help utilities to better address growing public concerns about PPCPs. The data collected in this study will help to identify those compounds resistant to removal and define critical SRT values where good removal is achieved.

Although no ecological or human health effects are known for most of the PPCP compounds in this study, effects may be discovered for these or other compounds in the future. This research is being conducted to improve our understanding of how theses micropollutants behave through secondary treatment so that we will have tools for addressing them should it become necessary in the future.

CHAPTER 2.0

MATERIALS AND METHODS

2.1 Project Approach

The present study was conducted to provide a better understanding of the passage of PPCPs through conventional municipal wastewater treatment systems. The experimental approach was based on the analysis of a target list of PPCPs with prior demonstrated occurrence in WWTP influents, to assess their relative removals from the aqueous phase of full-scale secondary treatment processes. This was accomplished through the analysis of samples collected from the influent and effluent streams from six conventional WWTPs and two pilot-scale MBRs located in southwestern United States. This region was ideal for this study because: 1) the presence of outdoor recreational personal care products such as sunscreens and mosquito repellents should be high because of the climate; 2) there is little heavy industry wastewater that could impose matrix interferences; 3) there is a great immediate need for this information, because of the high percentage of effluent-dominated streams and large demand for water recycling projects; and 4) the region provides accessibility to a variety of treatment facilities with similar influent wastewaters than might typically be available in other parts of the country. The plants encompass secondary ocean discharge, conventional tertiary discharge to effluent-dominated streams, and quaternary membrane treatment for water recycling, all within a very narrow regional area likely to have similar marketing distribution of household organic products.

Secondary treatment is typically defined as "the removal of biodegradable organic matter (in solution or suspension) and suspended solids" (Tchobanoglous et al., 2003). In a conventional activated sludge process, the aeration tank and the secondary clarifier form the secondary process unit (Clara et al., 2005). Compounds through secondary treatment are removed by biodegradation and solid adsorption. The purpose of this study was to evaluate the reduction of target compounds in the aqueous phase of the secondary influent and effluent. Therefore, this study does not define how much of the removal is caused by biodegradation versus solid adsorption.

2.1.1 Target Compounds

The selected target analytes are PPCPs that previously exhibited frequent occurrence in municipal WWTPs and are amenable to analysis by GC-MS-SIM (Loraine, 2004; Loraine and Pettigrove, 2002). Table 2-1 presents the target list of compounds, with their product classification and Chemical Abstracts Service (CAS) number.

Table 2-1. Selected Target Analytes.

Compound	CAS Number	Classification
Galaxolide	1222-05-5	Fragrance
Ethyl-3-phenylpropionate	2021-28-5	Fragrance
Methyl-3-phenylpropionate	103-25-3	Fragrance
Musk ketone	81-14-1	Fragrance
3-Phenylpropionate	501-52-0	Fragrance
Benzophenone	119-61-9	Sunscreen
Octylmethoxycinnamate	5466-77-3	Sunscreen
Oxybenzone	131-57-7	Sunscreen
Benzyl salicylate	118-58-1	Sunscreen
Ibuprofen	15687-27-1	Pharmaceutical
Caffeine	58-08-2	Pharmaceutical [1]
Triclosan	3380-34-5	Germicide
Chloroxylenol	88-04-0	Germicide
Tri(chloroethyl) phosphate (TCEP)	115-96-8	Fire retardant
Triphenylphosphate	115-86-6	Fire retardant
Methylparaben	99-76-3	Antioxidant
BHA	25013-16-5	Antioxidant
Butylbenzylphthalate	85-68-7	Plasticizer
Bis(2-ethylhexyl) phthalate	117-81-7	Plasticizer
DEET	134-62-3	Insect Repellent
Octylphenol	27193-28-8	Surfactant (detergent)

[1] Also from nonpharmaceutical sources

2.1.2 Participating Facilities

The participating facilities were selected to target a wide range of SRTs and treatment trains. The SRT conditions covered 0.5–30 days for facility capacities from 5 mgd to greater than 300 mgd. Whenever possible, sampling was performed in summer and winter months to investigate seasonal variations in plant performance. Unit process descriptions and secondary operating characteristics of these facilities are provided in Table 2-2, and process schematics for each WWTP are presented in Figures 2-1 to 2-6.

Plant A is a high-purity oxygen-activated sludge facility with an approximate SRT of 1 day. The wastewater treatment process consists of bar screening, chemical precipitation, primary clarification, oxygen generation, and secondary clarification. The majority of the secondary effluent is discharged to the ocean.

Plant B is a water reclamation plant with a secondary process SRT around 4 days. The process includes bar screening, primary sedimentation, aeration, secondary clarification, and media filtration. After demineralization and chlorination, the treated water is available for industrial purposes or irrigation. Only the secondary process was included in the study.

Plant C is an activated sludge facility with an SRT of around 5 days. It consists of screening, sedimentation, biological processes, filtration, and UV disinfection. A portion of the treated secondary effluent is discharged to an ocean outfall pipeline and the remainder is further treated for reclamation applications. Only the secondary process was included in the study.

Plant D is an advanced water reclamation plant with an SRT between 7 and 20 days. It consists of bar screening, primary treatment, an activated sludge system, granular media filtration, and microfiltration followed by reverse osmosis (RO). The effluent from the RO system has excellent water quality and can be potentially used for industrial reuse and groundwater recharge.

Plant E is an activated sludge facility with an SRT between 11 and 16 days. After coarse screening, grit removal, anoxic-aerobic biological nutrient removal, secondary clarification, and UV disinfection, the treated effluent is discharged to a nearby river. Plant E also provides treated water for reuse to local industries.

Plant F has an extended aeration process with an SRT between 20 and 30 days. It consists of bar screening, extended aeration, secondary clarification, media filtration, and UV disinfection. Treated water is discharged to a nearby wash.

Additionally, two pilot MBRs using different membrane configurations and operated at SRTs of 14–15 days were included as part of the sampling program. These MBRs are discussed in detail in Section 2.1.3.

Table 2-2. Descriptions of Participating Treatment Facilities.

Facility	Primary Treatment	Secondary Treatment	Secondary Aeration	MLSS[1] (mg/L)	SRT (days)	Filters	Disinfection
A	Polymer ferric	High- purity O_2-activated sludge	Pure O_2	1,300–2,600	0.5–1 5	None	None
B	No chemicals	MLE[2] Nit./Denit.	Diffused air	1,800–2,000	3–5	Deep bed[3]	Chlorine
C	No chemicals	Activated sludge	Diffused air	2,000–3,000	4-6	Deep bed[3]	UV
D	No chemicals	Nit./Denit.	Diffused air	2,500–3,000	7 – 20	Granular microfiltration/ RO	Chlorine
E	None	Nit./Denit.	Diffused air	2,100	11-16	None	UV
F	None	Ext. aeration Nit./Denit.	Surface Air	4,000	20–30	Deep bed[3]	UV
MBR #01[4,5]	n/a	Nit./Denit.	n/a	14,000	14	n/a	n/a
MBR #02[6,7]	n/a	n/a	n/a	11,500	15	n/a	n/a

[1] MLSS, mixed liquor suspended solids; [2] MLE, modified Ludzack Ettinger process; [3]Granular media; [4] Uses flat sheet membranes; [5] Located at plant E; [6] Uses free-end hollow fiber; [7] Located at a facility not listed above.

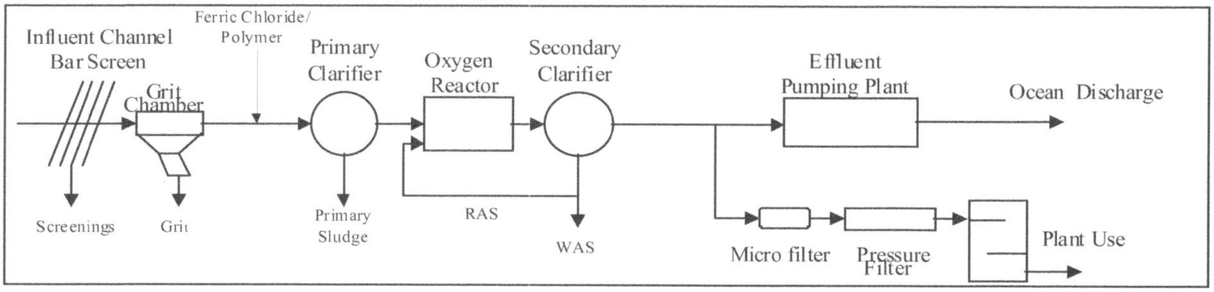

Figure 2-1. Schematic of Plant A (SRT ~ 0.5–1.5 days).

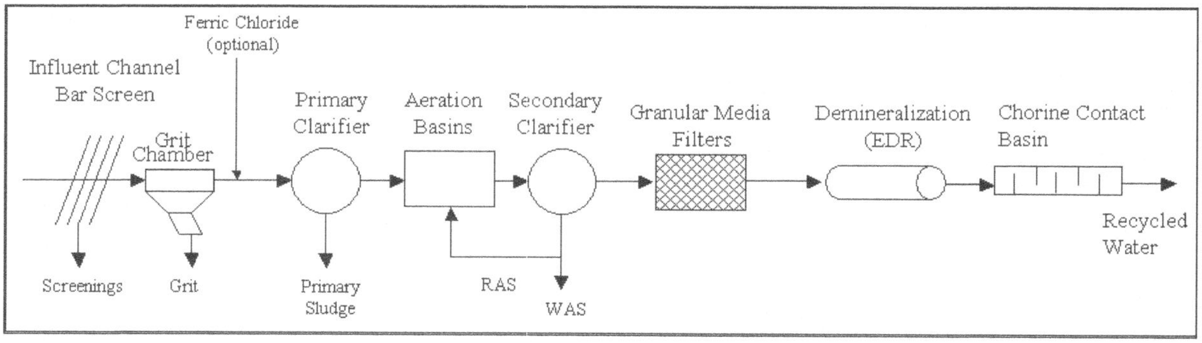

Figure 2-2. Schematic of Plant B (SRT ~ 3–5 days).

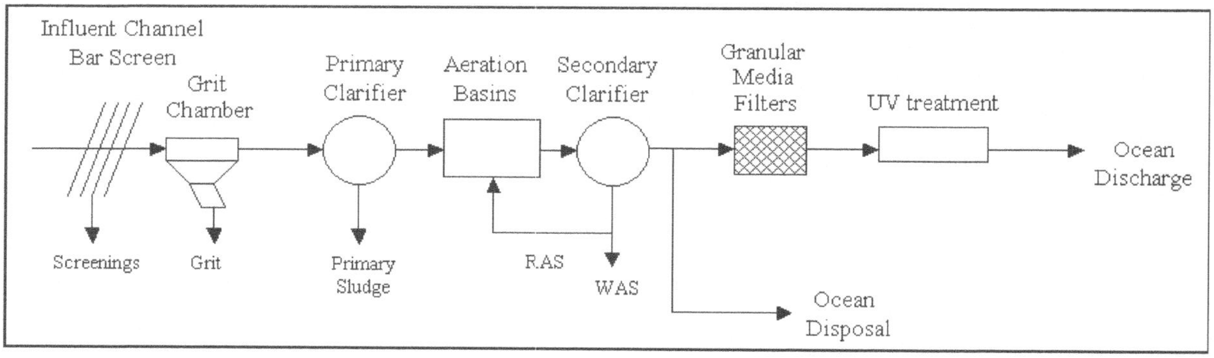

Figure 2-3. Schematic of Plant C (SRT ~ 5–6 days).

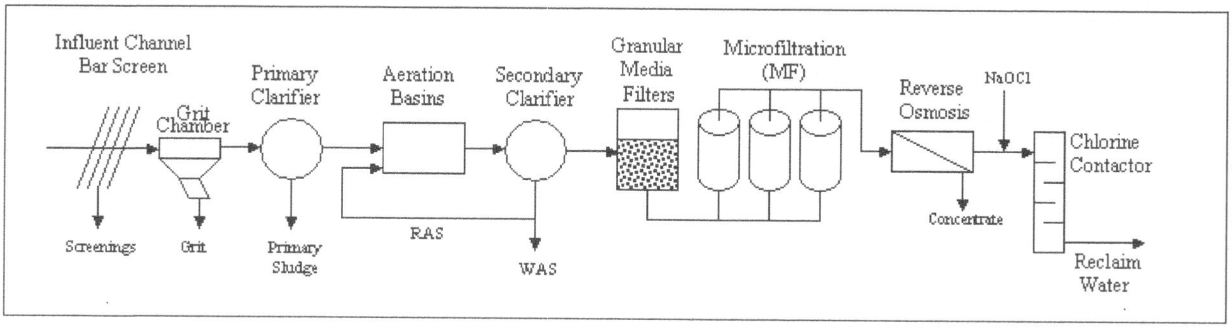

Figure 2-4. Schematic of Plant D (SRT ~ 7–20 days).

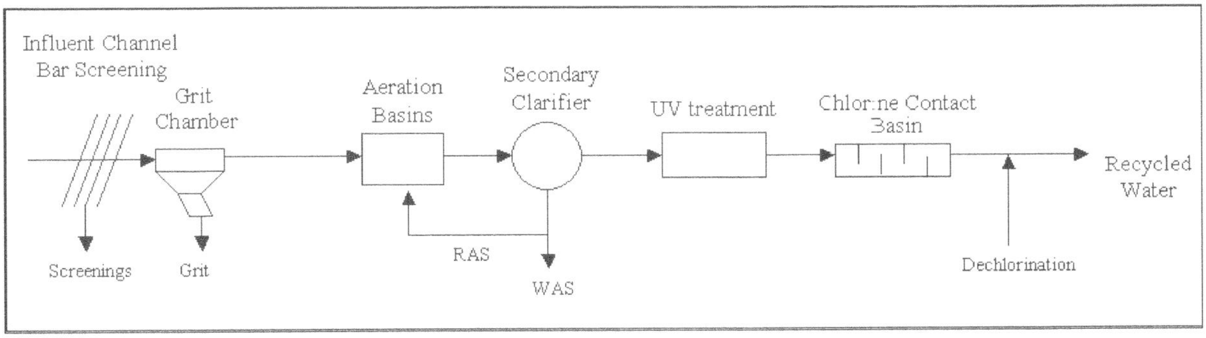

Figure 2 5. Schematic of Plant E (SRT ~ 11–16 days).

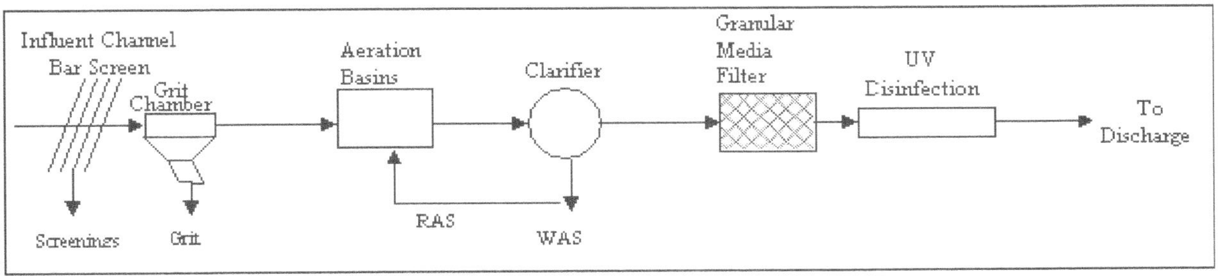

Figure 2-6. Schematic of Plant F (SRT ~ 25–30 days)

The sampling locations at each of the participant facilities are presented in Table 2-3. Samples of influent to secondary treatment were collected ahead of the return-activated sludge line.

Table 2-3. Sampling Locations at Participating Treatment Facilities.

Facility	Raw Influent	1° Effluent	2° Effluent	3° Effluent[1]	Advanced Treatment[2]
A		X	X		
B		X	X		
C		X	X		
D		X	X	X	X[2]
E	X[3]		X		
F	X[3]		X	X	
MBR #01	X[3]		X[4]		
MBR #02		X	X[4]		

[1] Samples of disinfected effluent were included when disinfection was applied; [2]Microfiltration/RO effluent; [3] Sampling had to be done here because of a lack of primary clarification. [4] MBR effluent.

2.1.3 Membrane Bioreactor Pilot Systems

Two MBR pilot systems were used in this study. Figure 2-7 provides a schematic of MBR #01 installed at Plant E. The design capacity of this pilot is 25 gallons per minute (gpm). This MBR is equipped with flat sheet membranes arranged in cassettes. In this unit, the feed water was passed through a traveling band screen before entering the feed holding tank. The feed water was then pumped to the denitrification zone using a submerged pump with a programmable logic controller. Water was then pumped to the prenitrification zone where it was aerated with fine bubble air. Mixed liquor flowed by gravity to the nitrification zone where filtration occurred. Constant coarse bubble aeration was provided in the nitrification zone to minimize fouling. Mixed liquor overflows back to the denitrification zone and sludge can be wasted daily from the prenitrification zone to maintain a constant sludge age. Sample collection occurred after a minimum of 2–3 SRTs when steady state was achieved (Trussell et al., 2005). Correct functioning of the MBR pilot was ensured by following customary QA/QC procedures recommended by the respective manufacturers.

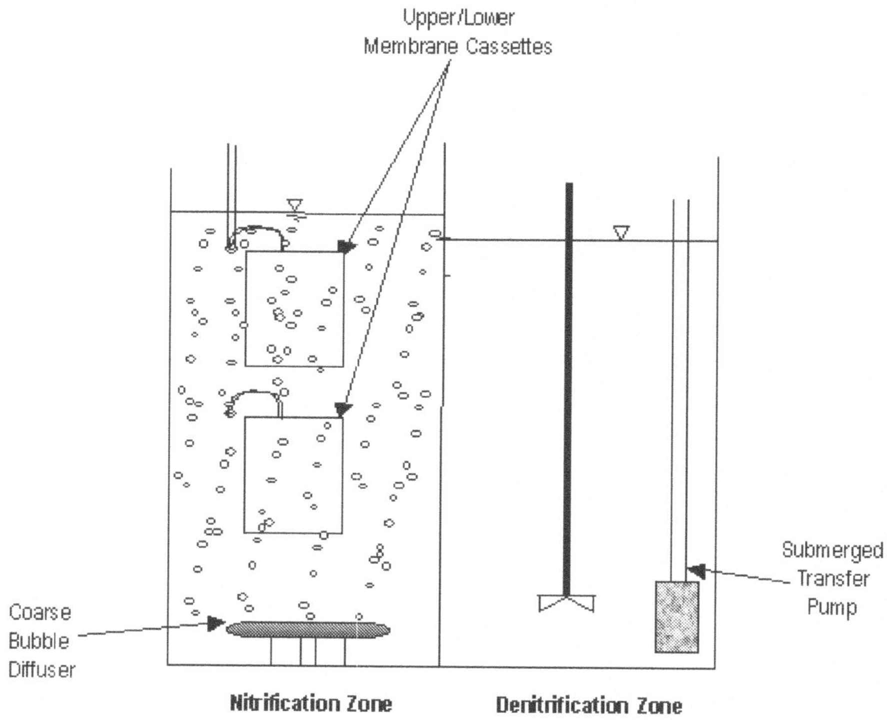

Figure 2-7. Schematic of the MBR Pilot #01.

Figure 2-8 presents a schematic of MBR #02 (15 gpm), which was installed at a WWTP with only primary treatment. The MBR pilot unit consists of an aerobic tank, an anoxic tank, and a membrane tank equipped with free-end hollow fiber membranes. Feed water to the system is screened before being passed to the anoxic basin. The anoxic basin is designed with fine bubble diffusers so that it can be used either as an aerobic or anoxic zone. Water from the anoxic tank flows by gravity to the aeration tank for nitrification. Nitrified water from the aeration tank is recirculated back to the anoxic tank for denitrification. Water from the aeration tank is also recirculated to the membrane tank at a fixed flowrate, and the overflow from the membrane tank flows back to the aeration tank by gravity. Wastewater could be nitrified and denitrified before being filtered out from the membrane tank. Sludge wasting is done automatically from the aeration tank and it is controlled via programmable logic controller after receiving an output from a TSS sensor submerged in the aeration tank.

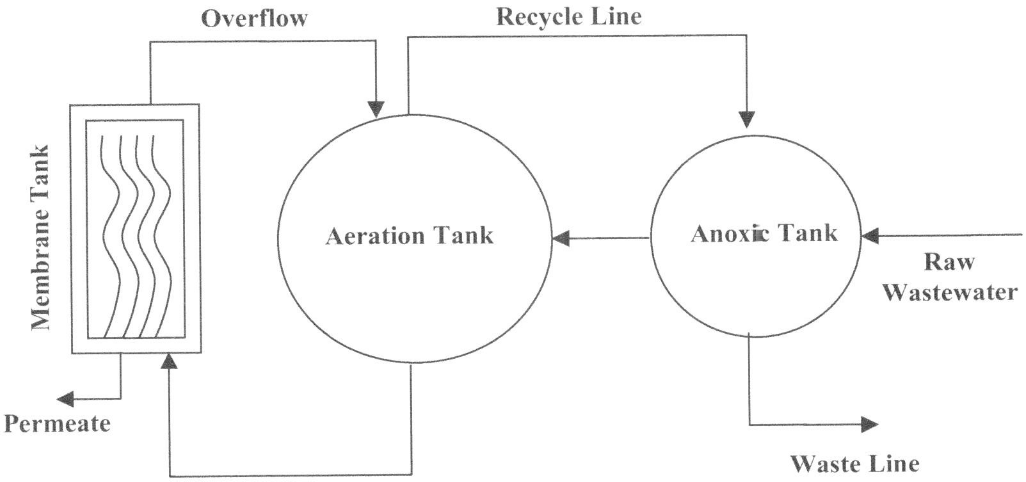

Figure 2-8. Schematic of the MBR Pilot #02.

2.2 Sampling Protocols

Sampling protocols were designed to ensure that all collected samples were representative of typical plant operating conditions. The 24-h time-weighted composite samples were collected instead of flow-composite samples, because of the absence of flow measurement devices at all of the specified sampling locations. The validity of this approach is supported by comparisons of time-weighted and flow-weighted composite sampling results for BOD and TSS that were demonstrated to fall within 10% of each other (Metcalf & Eddy, 2003).

All sampling events were conducted during periods of normal plant operation. No efforts were made to modify any standard plant operating procedures. Sampling was performed using dedicated Glacier® Composite samplers. The technical data sheet for this equipment, manufactured by Teledyne ISCO (Lincoln, NE.), is presented in Appendix A. To secure the quality of the data generated for the project, three samplers were used for sample collection at each facility (i.e., one for each potential sampling location). Each composite sample had a total volume of approximately 10 L. From this amount, approximately 2 L were transferred to amber glass sample bottles with Teflon-lined caps and submitted to the analytical laboratory for PPCP analysis. One blank and one duplicate split sample were collected per plant sampling event. The blank sample collection procedure consisted of running laboratory-grade-distilled water through one of the sampling devices during each sampling event, before any sample collection. This water was then transferred to the appropriate sampling bottles and constituted the blank sample for each sampling event. Of the nineteen blank samples generated per compound, all were nondetect except for benzophenone (one occurrence 10% above MDL), oxybenzone (one occurrence 100% above MDL), triphenylphosphate (one occurrence 30% above MDL), and butylbenzylphthalate (three occurrences 10–20% above MDL). The duplicate split sample location was rotated during subsequent seasonal sampling of each facility. The majority of the split samples fell within a relative percent difference (RPD) of 35% with a few sporadic exceptions. No blank or split corrections were made to the data. The remaining sample volume was aliquoted into appropriate bottles to conduct analyses for conventional constituents [TSS, volatile suspended solids (VSS), total Kjeldahl nitrogen (TKN), and NO_3–N, NO_2–N, and NH_3–N] using standard methods. A certified laboratory (Applied Physics and Chemistry Laboratory, Chino, Calif.) conducted these analyses. TSS and VSS data were only used when onsite plant data for those parameters were not available. Details of the sampling protocol used for collecting 24-h weighted composite samples, including shipping and handling methods is presented in Appendix B.

After completing every sampling event, operational data (current and historical) were collected. These data were used to assess any apparent relationship between effluent concentrations or removal percentages of PPCPs and operational SRTs observed from collected samples. Plant data were collected in the following categories:

- Plant identification
- Flow
- Unit processes (with schematic)
- Pretreatment program
- Water quality for plant influent and effluent
- Primary process operation data during sampling
- Secondary process operation data during sampling
- Tertiary process operation data during sampling
- Membrane treatment operation data during sampling
- Disinfection operation data during sampling
- Sampling protocol information

A comparison of the sampling-day values and historical average values of these parameters was made to confirm normal plant operation on the sampling dates. This information was provided by the plant personnel and collected using custom-designed data log sheets, which are presented in Appendix C. Historical averages of plant operational parameters and water quality indicated that plant operation was normal during the sampling days. This information is presented in Appendix D. Finally, additional details on the overall QA/QC procedures used for conducting these tasks are presented in Appendix E.

2.3 Analytical Protocols

The samples collected for this project were processed using filtration, solid phase extraction (SPE), and further analyzed by GC-MS by Dr. Greg Loraine, formerly with San Diego State University and currently with Dynaflow, Inc. (Jessup, Md.). A Perkin Elmer Clarus 500 GC-MS, which provides simultaneous collection of total ion count and selective ion monitoring (SIM) data, was used. Dynaflow provided all required GC-MS-SIM analyses of wastewater samples, calibrations, and reanalysis of extracted samples.

Sample Handling and Extraction Procedures

All samples were refrigerated upon receipt and 2 L were extracted in most cases within 24 h. Wastewater samples were prefiltered through precleaned glass fiber filters (a Fisherbrand G8, 2.5-mm particle retention, and a Fisherbrand toxicity characteristic leaching procedure filter, 0.7-mm pore size; Fisher Scientific, Tustin, Calif.). The filtrate was acidified with concentrated sulfuric acid to a pH <2.0. After acidification, aqueous samples were extracted using sulfonated polystyrene divinylbenzene solid phase extraction disks (3M Empore SDB-RPS; 3M, Minneapolis, Minn.). Before use, the disks were precleaned with ethyl acetate and then conditioned per manufacturer's recommendation. The flowrate through the disk was maintained at less than 50 mL/min. The SPE disks were then eluted with acetone (3 mL) and a triple aliquot of ethyl acetate (5 mL). The eluate was dried over Na_2SO_4 and the volume reduced to 0.5 mL. This method is similar to those used by Ollers et al. (2001) and Jonsson et al. (2002). The concentrate was transferred to a 2-mL GC vial sealed with screw caps with PTFE-coated rubber septa. 4,4-Dibromooctafluorobiphenyl (Supelco, Bellafonte, Pa.) was added before analysis as an internal standard.

Analytical Method

The GC method was as follows: 2.0-µL injection, splitless, injector temperature 175°C, 40°C for 2 min, then 10°C/min ramp to 250 °C, hold for 14 min, ramp to 290°C at a rate of 25°C/min, hold for 2 min. The column was a RTX-MS5 30 m x 0.25 mm ID, 0.25-µm film thickness (Restek). The Clarus 500 GC-MS was run in total ion count mode and select ion reporting (SIR) mode during the same run. The total ion count masses were scanned from *m/z* 70–350 from 6–37 min. There were five SIR segments. SIR 1 scans *m/z* 150, 156, 164, and 178 from 12.1 to 14.0 min. SIR 2 scans *m/z* 152, 163, 165, 182, and 190 from 14.0 to 17.2 min. SIR 3 scans *m/z* 194, 197, 206, 228, 243, and 249 from 17.5 to 20.0 min. SIR 4 scans *m/z* 178, 227, 279, and 290 from 20.0 to 23.0 min, and SIR 5 scans *m/z* 149, 206, and 326 from 23.0 to 26.5 min. Retention times and monitored ions are listed in Table 2-4.

Materials

The phthalates standards were prepared using an EPA 525 Update Phthalate Ester Mix (500 µg/mL; Supelco, Bellafonte, Pa). The internal standard, 4,4-dibromooctafluorobiphenyl, was also obtained from Supelco. Ethyl 3-phenylpropionate (98%), benzophenone (99%), caffeine (98.5%), benzyl salicylate (98%), tris(2-chloroethyl) phosphate (97%), BHA (96%), 3-phenylpropionate (99%), oxybenzone (98%), and octylmethoxycinnamate (98%) were obtained from Acros (Allentown, Pa). Galaxolide (50% in diethyl phthalate), 4-octylphenol (99%), musk ketone (98%), methylparaben (National Formulary/Food Chemical Codex), chloroxylenol (99%)

Table 2-4. Retention Times, Major Ions for Each Analyte, SIR Segment Used for Quantification.

Compound	Retention Time (min)	Major Ion m/z	Segment
Methyl-3-phenylpropionate	12.2	164	1
3-Phenylpropionate	13.2	150	1
Ethyl-3-phenylpropionate	13.2	178	1
Chloroxylenol	13.7	156	1
Methylparaben	14.7	152	2
BHA	14.9	165	2
DEET	16.1	190	2
Ibuprofen	16.5	163	2
Benzophenone	16.8	182	2
Dibromooctafluorebiphenyl – internal standard	17.3		TIC
TCEP	18.1	249	3
Octylphenol	18.2	206	3
Caffeine	19.1	194	3
Galaxolide	19.2	243	3
Benzyl salicylate	19.4	228	3
Musk ketone	20.4	279	4
Oxybenzone	20.9	227	4
Triclosan	21.7	290	4
Octylmethoxycinnamate	22.1	178	4
Butylbenzyl phthalate	23.7	206	5
Triphenylphosphate	24.3	326	5
Bis(2-ethylhexyl) phthalate	25.9	149	5
2,4-Dichlorophenol-d_3 – surrogate	10.7	165	TIC
Caffeine-$^{13}C_3$ – surrogate	19.1	197	3
3,3-Dichlorobenzidine-d_6 – surrogate	25.2	258	TIC
Di(ethylhexyl) phthalate-d_4 – surrogate	26.1	283	TIC

were obtained from Aldrich (St. Louis, Mo.). Methyl-3-phenylpropionate (98%) was obtained from Alfa Aesar (Wardhill, Mass.). Triclosan (Irgasan 97%; Fluka, Ronkonkoma, N.Y.), triphenylphosphate (500 µg/mL; Protocol), ibuprofen (98%, Sigma), and DEET (100 µg/mL; Ultrascientific, North Kingstown, R.I.) were also used. Radio-labeled surrogates, 2,4-dichlorophenol-ring D_3, (98%), bis(2-ethylhexyl)phthalate-ring-d_4, (98%), 3,3'-dichlorobenzidine-diphenyl-D_6 (98%), and caffeine-trimethyl-$^{13}C_3$, 99%) were obtained from Cambridge Isotope Laboratories (Andover, Mass.).

Calibration

Table 2-5 presents the root mean square values obtained for calibration curves prepared for each of the analyzed compounds.

Calibration curves were prepared using four concentration levels spanning at least one order of magnitude. All valid calibration curves had a root mean square of at least 0.95. The calibration standards were prepared to cover the range of expected concentrations after extraction and concentration from ~ 2000 mL to 0.5 mL. Thus, the instrument detection limits (IDLs) were reported in µg/mL. The IDLs were determined for pure compounds in ethyl acetate in accordance with U.S. Environmental Protection Agency guidelines (U.S. EPA, 1993).

Table 2-5. IDLs and Calibration Curve Root Mean Square Values for Each Analyte.

Compound	IDL (µg/mL)	Sample Concentration (µg/mL)	R^2
Butylbenzyl phthalate	0.030	0.025	0.97
Bis(2-ethylhexyl) phthalate	0.079	0.05	0.99
Methyl-3-phenylpropionate	0.064	0.05	0.99
Ethyl-3-phenylpropionate	0.016	0.05	0.99
3-Phenylpropionate	0.39	0.25	0.97
Benzophenone	0.013	0.025	0.99
2-Octyl-4-methoxycinnamate	0.039	0.025	0.97
Oxybenzone	0.24	0.05	0.95
Benzyl salicylate	0.40	0.05	0.97
Ibuprofen	3.6	1.0	0.96
Caffeine	0.025	0.025	0.98
Triclosan	0.10	0.05	0.96
Chloroxylenol	0.017	0.025	0.96
Galaxolide	0.011	0.025	0.99
Musk ketone	0.019	0.025	0.98
Methylparaben	0.34	0.1	0.97
BHA	0.034	0.25	0.99
DEET	0.090	0.25	0.99
TCEP	0.051	0.01	0.99
Triphenylphosphate	0.86	0.5	0.98
Octylphenol	0.037	0.1	0.98

Blanks

The extraction method was also performed on several aliquots of laboratory grade distilled water. In previous experiments several compounds were identified in the water blanks, including 3-phenylpropionate, benzophenone, triphenylphosphate, butylbenzyl phthalate, and bis(2-ethylhexyl) phthalate. All were below the IDLs with the exception of bis(2-ethylhexyl) phthalate, which was 4–5 µg/L. The major source of the phthalic acid esters was determined to be the SPE disks (Loraine & Pettigrove, 2002). For this reason, the method was modified to include a step where the SPE disks were washed with three aliquots of ethyl acetate before conditioning. Although this step reduces the level of contamination, the concentrations of bis(2-ethylhexyl) phthalate remained high. The U.S. EPA guidance specifies that a compound cannot be reported if it is found in the field blank, unless its concentration is 10 times greater than that of the blanks (Till, 2003). According to this rule, the reporting limit for bis(2-ethylhexyl) phthalate reached 48 µg/L, and for this reason, this compound was excluded from the analysis. The target list was therefore reduced to 20 compounds.

Method Detection Limits

Method detection limits (MDLs) for each compound presented in the fourth column of Table 2-6 were determined in accordance with U.S. EPA guidelines (U. S. EPA, 1993) using a wastewater secondary effluent. The method was as follows: Approximately five gallons (19 L) of effluent were collected as a grab sample. The effluent was filtered using dual glass fiber filters and stored at 5°C. The refrigerator temperature was lowered to 4°C shortly after. Two liters of

Table 2-6. Method Detection Limits and Ratios of the MDL to the Initial Spiked Concentrations for Each Analyte.

Compound	Spiked Conc., µg/L	Baseline Conc., [1] µg/L	MDL, µg/L	Ratio of MDL to Initial Conc.
Butylbenzyl phthalate	0.01	0.142	0.13	0.88
Bis(2-ethylhexyl) phthalate [2]	0.01	5.37	5.03	0.93
Methyl-3-phenylpropionate	0.05	BDL	0.043	0.86
Ethyl-3-phenylpropionate	0.05	BDL	0.012	0.25
3-Phenylpropionate	0.5	2.43	10.7	3.64
Benzophenone	0.05	0.112	0.043	0.27
Octylmethoxycinnamate	0.01	0.074	0.11	1.28
Oxybenzone	0.01	0.083	0.06	0.65
Benzyl salicylate	0.05	BDL	0.022	0.45
Ibuprofen	1.0	BDL	1.44	1.44
Caffeine	0.05	0.019	0.053	1.06
Triclosan	0.05	0.19	0.35	1.44
Chloroxylenol	0.05	BDL	0.015	0.29
Galaxolide	0.01	1.58	0.61	0.39
Musk ketone	0.01	0.065	0.034	0.45
Methylparaben	0.5	BDL	0.30	0.61
BHA	0.05	BDL	0.011	0.23
DEET	0.05	0.189	0.086	0.36
TCEP	0.05	0.302	0.179	0.51
Triphenylphosphate	0.5	BDL	0.62	1.24
Octylphenol	0.01	BDL	0.004	0.31

[1]BDL = below detection limit; [2] Reporting limit for bis(2-ethylhexyl) phthalate is 48 µg/L, as stated in Section 2.3.2 "blanks."

sample were extracted and analyzed as a baseline for the background concentrations. The remaining sample was divided into seven 2-L aliquots. Each aliquot was spiked with a stock solution containing the surrogates and a mixture of the analytes, according to the list presented in the first column of Table 2-6. The second column of this table shows the spiked concentrations of each analyte, and the third column shows the compounds of interest in the wastewater matrix before spiking (i.e., baseline concentrations), as observed in previous experiments. The retention times and main ions for each analyte are listed in Table 2-4. Four surrogates were used, including caffeine-$^{13}C_3$, 3,3-dichlorobenzidine-d_3, 2,4-dichlorophenol-d_3, and bis(ethylhexyl)phthalate-d_4 (Cambridge Isotope Laboratories, Andover, Mass.). This combination of surrogates provided a range of polarities and acidities similar to the analytes. The selected spiked concentrations were 0.25, 1.25, 0.25, and 2.5 µg/L, respectively. Because surrogate recoveries must be greater than 70% and less than 130%, if any of the surrogates failed these criteria, the samples were reanalyzed. The resulting MDLs (µg/L) are listed in the fourth column of Table 2-6. The concentrations used for the MDL studies were no more than five times the MDL determined for all of the compounds, as can be seen in the last column of this table, which shows the ratio of MDL to initial concentration for each analyte. If the baseline concentration of a given analyte was 20% or greater than its spiked concentration, the ratio of MDL to initial concentration was calculated from the total of the spiked and baseline concentrations.

As GC-MS is not the method that provides the lowest possible MDL, we did review the procedures for the MDL calculation to verify reasonable estimates of the PPCP concentrations. One concern was background concentration of the compounds present in the effluent matrix used for the detection limit assessment. According to U.S. EPA guidelines (U.S. EPA, 1993), the

MDL can be determined in a sample matrix other than reagent water, and if the measured level of analyte is greater than five times the estimated detection limit, the sample may be used as is to determine the MDL if the analyte does not exceed 10 times the MDL of the analyte in reagent water. Among the 20 target PPCPs, butylbenzyl phthalate, ethyl-3-phenylpropionate, 3-phenylpropionate, benzophenone, chloroxylenol, Galaxolide, musk ketone, and TCEP do not meet this requirement. For these compounds, we further reviewed the RPD obtained for the split samples collected during each sampling event. All the split samples showed sufficient precision (i.e., RPD within 20–30%) except for ethyl-3-phenylpropionate (Appendix F). This demonstrates that for the majority of the target analytes appropriate MDL values have been set. Ethyl-3-phenylpropionate was not frequently detected in the samples and the precision was indeterminate. If the background concentration is too high for ethyl-3-phenylpropionate, the MDL might be estimated to be unreasonably low.

CHAPTER 3.0

RESULTS AND DISCUSSION

The PPCP data were analyzed to identify trends in their occurrence in wastewater influent (i.e., influent to the bioprocess, which could be preliminary treated effluent or primary treated effluent, depending on the facility) and to characterize their removal from the aqueous phase of secondary treatment.

The occurrence of PPCPs in wastewater influent was characterized by the following evaluations:

- The influent concentration distributions observed for each target compound
- The distribution of the target compounds into three discrete occurrence bin classifications

The removal of PPCPs during secondary biological treatment was characterized by the following evaluations:

- The percent removals achieved for each target compound at each facility for each sampling event,
- The distribution of the target compounds into three discrete treatment bin classifications,
- An evaluation of the relationship of compound removal to treatment SRT,
- The identification of a minimum SRT level to achieve good removal for each target compound,
- A comparison of the removal performance of conventional activated sludge processes and MBRs

Additionally, a series of qualitative observations are provided. These observations lack sufficient data to allow definitive conclusions to be drawn, but they are suggestive of trends that might be worthy of additional study.

3.1 Occurrence of Soluble PPCPs in Sampled WWTPs

The occurrence of target PPCP compounds in the secondary influent samples were characterized in terms of cumulative probability against concentration. The PPCP compounds were then classified into three "occurrence bins," based on their observed frequency distribution, as defined in Table 3-1.

Table 3-1. Definition of Occurrence Bins.

Bin Number	Description	Bin Assignment Criteria
O1	Infrequent occurrence	Detected in less than 25% of the observations in influent
O2	Intermediate occurrence	Detected between 20% and 75% of the observations in influent
O3	Frequent occurrence	Detected in more than 75% of the observations in influent

The probability plots used to classify the compounds into the occurrence bins are presented in Figures 3-1 through 3-3 (Bins O1, O2, and O3, respectively). For each of the analyzed PPCPs, the probability plots indicate the percentage of cases when their concentration was equal or below a given value. Values reported as "below MDL" are plotted at the MDL and shown in open symbols; measured concentrations are shown in solid symbols. The concentration data for bis(2-ethylhexyl) phthalate was not included because it has proven to be ubiquitous to the sampling and analytical processes (Fromme et al., 2002) and was excluded from the original targeted list (see Section 2.3). Therefore, this study monitored 20 PPCPs.

As shown in Figure 3-1, TCEP, octylphenol, methyl-3-phenylpropionate, and triphenlyphosphate were classified as part of Bin O1 (i.e., infrequent occurrence), given that their concentration was measurable in less than 25% of the observations. This indicates that it was quite infrequent for these compounds to occur in the influent to activated sludge processes at concentrations higher than the MDL.

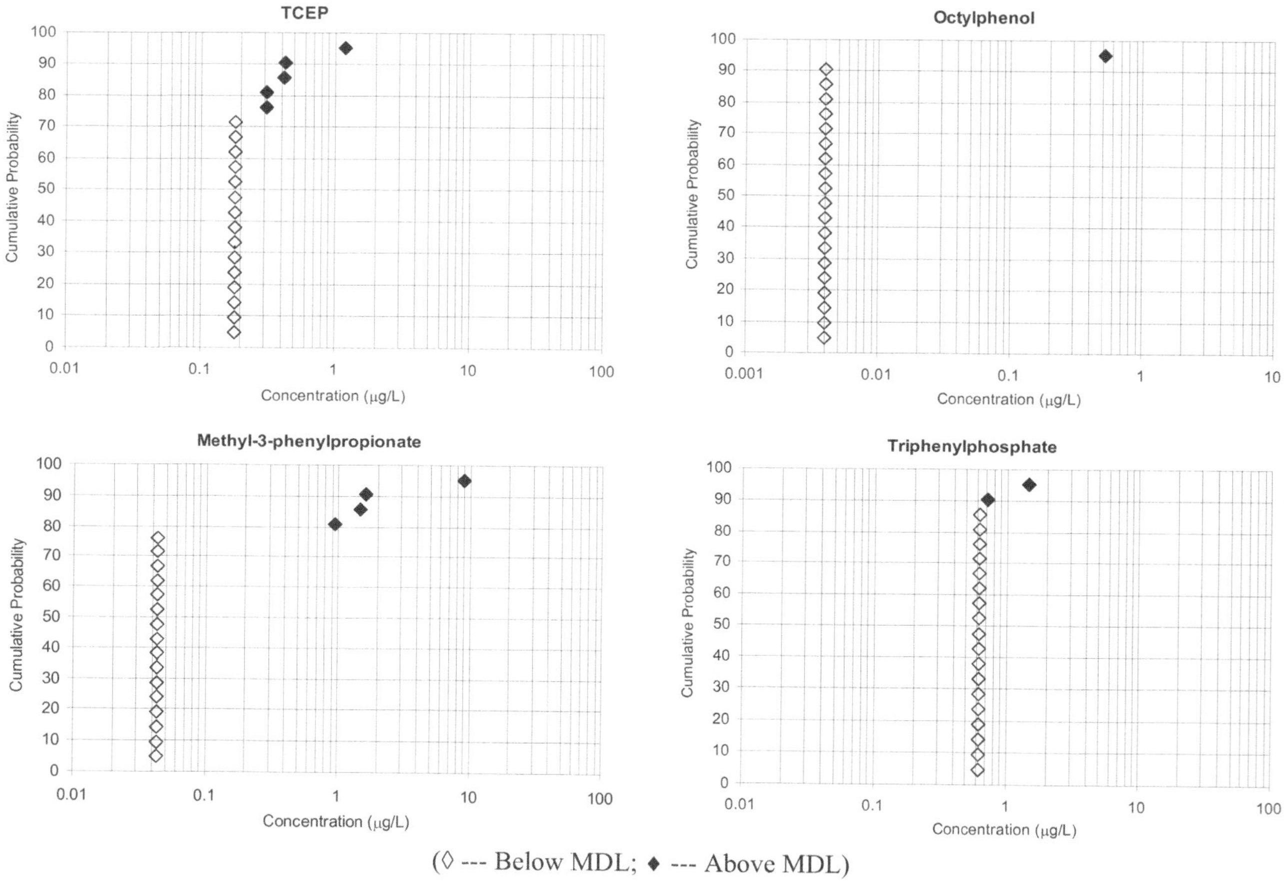

Figure 3-1. Cumulative Probability Distribution for the Occurrence of Bin O1 (Infrequent Occurrence) Compounds in Influent to Secondary Treatment.

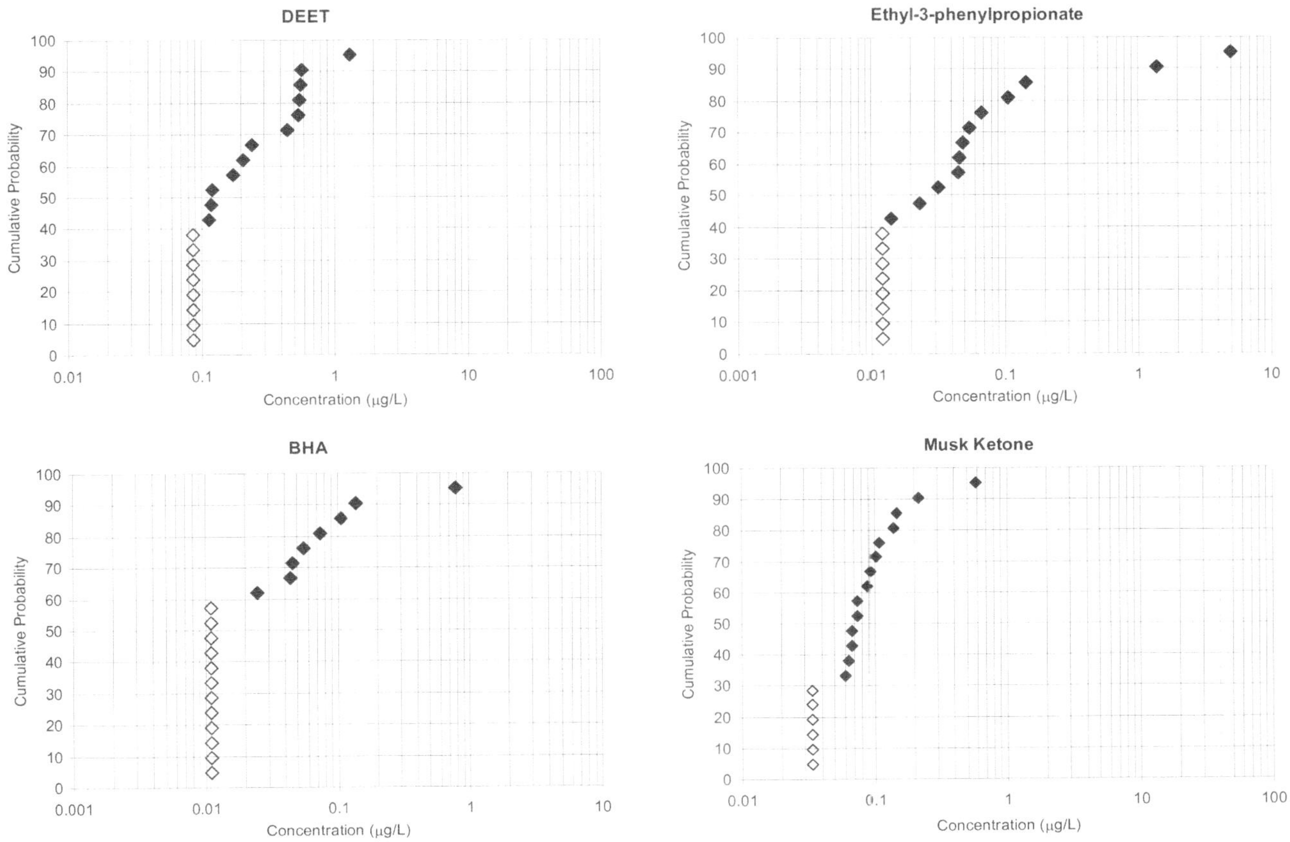

(◊ --- Below MDL; ♦ --- Above MDL)

Figure 3-2. Cumulative Probability Distribution for the Occurrence of Bin O2 (Intermediate Occurrence) Compounds in Influent to Secondary Treatment.

In contrast, the compounds presented in Figure 3-2 were more frequently detected, and were included as part of Bin O2 (i.e., intermediate occurrence; measurable concentrations in 25–75% of the cases). Quantifiable concentrations (i.e., above the MDL) occurred in approximately 40% of the cases for BHA, 60% of the cases for DEET and ethyl 3-phenylpropionate and 70% of the cases for musk ketone.

The remaining compounds were classified within Bin O3 (frequent occurrence), as their concentrations were above the MDL in greater than 75% of the cases. Bin 3 included the majority of compounds from the target list, as presented in Figure 3-3.

Figures 3-1 through 3-3 show that 20% of the target compounds were classified in Bin O1 (i.e., infrequent occurrence), 20% in Bin O2 (i.e., intermediate occurrence), and 60% in Bin 3 (i.e., frequent occurrence). The observed PPCP concentrations at the 10th, 25th, 50th, 75th, and 95th percentile of cumulative occurrence are presented in Table 3-2. These values indicate that the 50th percentile concentration of the monitored PPCPs in secondary influent were trace levels (i.e., median concentration < 10 µg/L) except for 3-phenylpropionate.

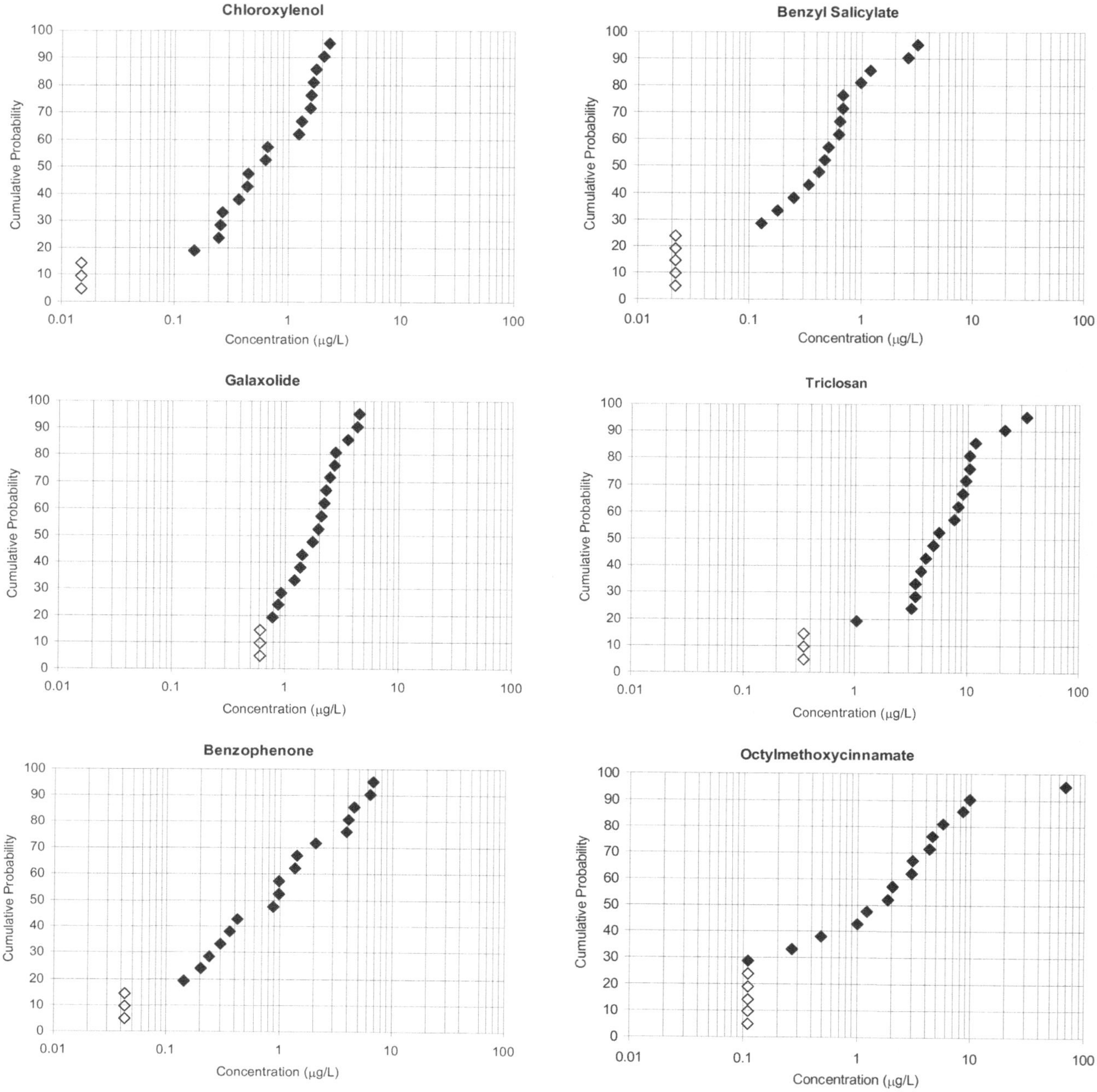

(◊ --- Below MDL; ♦ --- Above MDL)

Figure 3-3. Cumulative Probability Distribution for the Occurrence of Bin O3 (Frequent Occurrence) Compounds in Influent to Secondary Treatment.

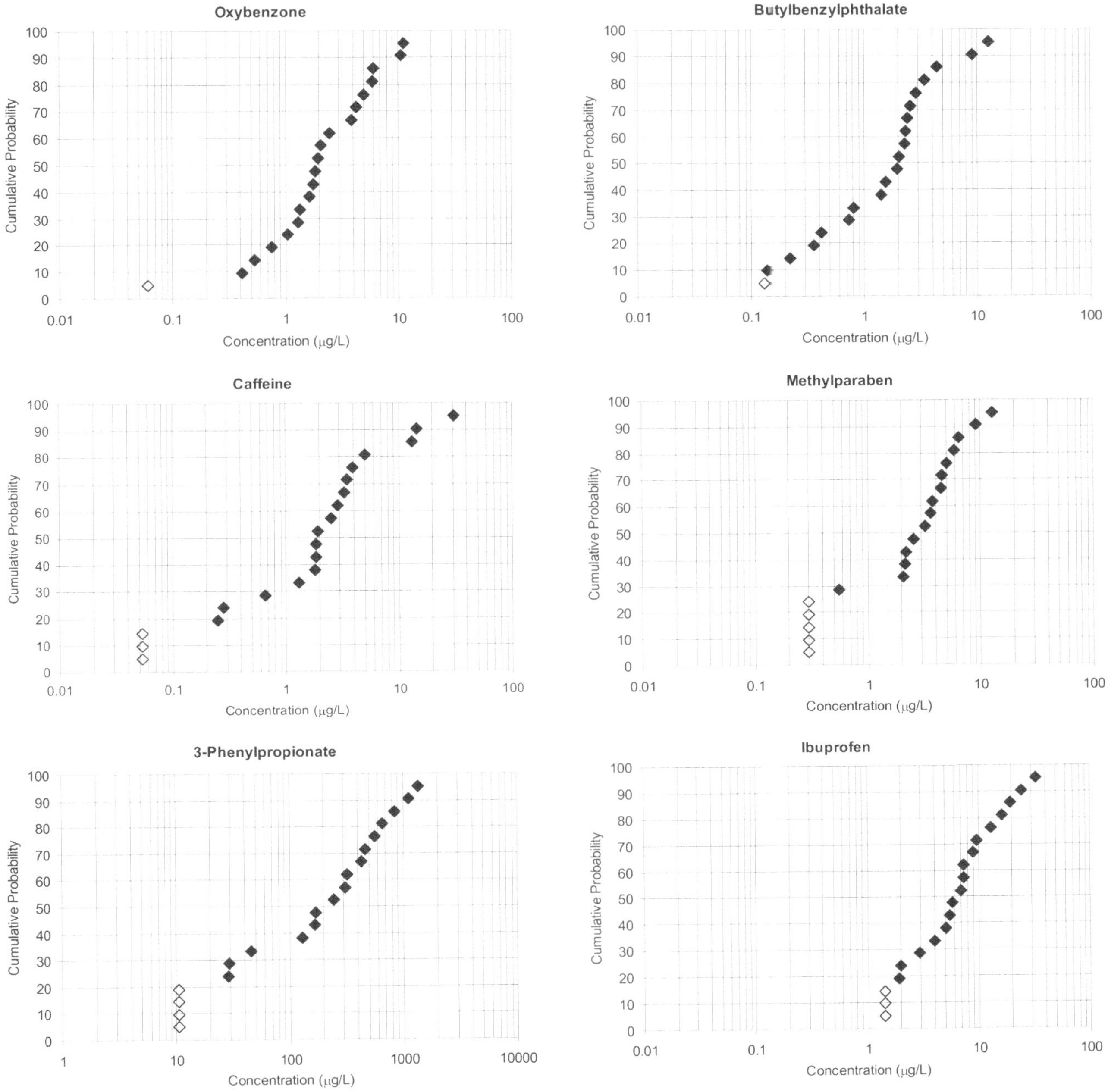

(◊ --- Below MDL; ♦ --- Above MDL)

Figure 3-3 (continued). Cumulative Probability Distribution for the Occurrence of Bin O3 (Frequent Occurrence) Compounds in Influent to Secondary Treatment.

Table 3-2. Concentrations of PPCPs at Different Occurrence Percentiles.

Bins	Compounds	Concentration (μg/L) at percentiles				
		10th	25th	50th	75th	95th
Bin O1 Infrequent	TCEP	ND	ND	ND	ND	1.2
	Octylphenol	ND	ND	ND	ND	0.51
	Triphenylphosphate	ND	ND	ND	ND	1.4
	Methyl-3-phenylpropionate	ND	ND	ND	ND	9.0
Bin O2 Intermediate	BHA	ND	ND	ND	0.052	0.79
	DEET	ND	ND	0.12	0.54	1.3
	Musk ketone	ND	ND	0.070	0.11	0.58
	Ethyl-3-phenylpropionate	ND	ND	0.026	0.062	5.0
Bin O3 Frequent	Methylparaben	ND	0.033	3.0	5.0	13
	Benzyl salicylate	ND	0.050	0.45	0.68	3.2
	Chloroxylenol	ND	0.25	0.52	1.6	2.3
	Triclosan	ND	3.3	5.2	10	34
	Octylmethoxycinnamate	ND	0.11	1.4	4.6	68
	Butylbenzyl phthalate	0.15	0.50	2.0	2.8	13
	Benzophenone	ND	0.21	0.94	3.3	6.7
	Galaxolide	ND	0.90	1.9	2.6	4.5
	Ibuprofen	ND	2.1	6.3	12	32
	Oxybenzone	0.42	1.1	1.9	4.6	11
	Caffeine	ND	0.32	1.9	3.8	31
	3-Phenylpropionate	ND	29	205	530	1380

3.2 Removal of Soluble PPCPs by Secondary Wastewater Treatment

3.2.1 Plant Operation and Performance Data

Operational and performance data of each of the sampled facilities were compiled. As presented in Appendix D, the operational parameters included SRT, hydraulic retention time (HRT), MLSS, and return-activated sludge flowrate as a percentage of the influent flowrate. This information highlights the diversity of the group of facilities included in this study. Monthly averages of the operation data are also presented to demonstrate that the sampling events occurred during periods of typical plant operation.

Plant performance data (secondary influent and effluent characteristics) were collected for each sampling event (i.e., also provided in Appendix D). This information gives an indication of plant performance at the time of the sampling. The parameters monitored by the facilities include BOD, TSS, and VSS. Monthly averages of these values were also collected with the objective of establishing comparisons with the values reported during the sampling events. In addition, Applied P&CH Laboratories (Chino, Calif.) analyzed the TKN, ammonia as nitrogen (NH_3–N), nitrite as nitrogen (NO_2–N), and nitrate as nitrogen (NO_3–N) in influent and effluent samples.

The compiled operation and performance data generally demonstrated that all sampled facilities were performing within normal limits during the sampling events.

3.2.2 Removal of Target PPCPs through Secondary Treatment

The efficiency of secondary treatment for preventing the passage of the target PPCPs was evaluated by comparing the concentration of each compound in secondary influent and effluent. The percentage removal (P) of each compound through secondary treatment was calculated as:

$$P = [(\text{Influent} - \text{Effluent})/(\text{Influent})] * 100 \qquad [1]$$

Following the same approach employed to examine PPCP occurrence, a "treatment bin" classification was developed to characterize the target PPCPs by their degree of removal through secondary treatment. The criteria for defining these treatment bins are presented in Figure 3-3.

Table 3-3. Definition of Treatment Bins.

Bin Number	Description	Bin Assignment Criteria
T1	Good removal	Median of the percent removal values is greater than 80%
T2	Moderate removal	Median of the percent removal values is between 50% and 80%
T3	Poor removal	Median of the percent removal values is less than 50%

The percent removal data for each analyzed PPCP, classified by plant and sampling event, is presented in Table 3-4. The corresponding treatment bin classification for each PPCP compound as defined above is also indicated.

Table 3-4. Percent Removal and Treatment Bins of Target PPCP Compounds.[1,2,3]

PPCP Compound	Plant A	Plant B	Plant C	Plant D	Plant E	Plant F	MBR #01	MBR #02	Bin
3-Phenylpropionate	>92[1] >98 97	>77 ND >63	ND >94 >94	>64 >98 >99	ND[2] >97 --	>99 >96 ND	>99 >97 >98	ND >97 --	1
Octylmethoxycinnamate	>85 >94 >95	20 57 ND	>97 >99 ND	>96 >96 ND	>89 >99 --	>99.8 >98 ND	>98 ND ND	>89 >99 --	1
Methylparaben	>42 >93 98	ND ND ND	>92 >95 >88	>86 >94 >92	ND >93 --	>98 ND >86	>95 >86 >97	ND >93 --	1
Caffeine	0*[3] 79 91	ND ND ND	>97 >99.6 >96	>92 >97 >98	>79 >98 --	>99.8 >99 >98	>99.9 >98 >99	>79 >98 --	1
Ibuprofen	ND >84 >75	43 0* 0*	>80 >89 >74	>64 ND >80	ND >85 --	>94 >79 >72	>96 >93 >91	ND >85 --	1
Chloroxylenol	53 >72 77	15 0* 0*	90 90 >98	ND 90 ND	ND >98 --	99 93 >96	>99 >99 >99	0* >98 --	1
Benzyl salicylate	28 >95 94	ND ND ND	>97 >99 >97	>97 >95 >96	>91 >97 --	>99 >98 >88	ND ND >98	>91 >97 --	1

Table 3-4 (continued). Percent Removal and Treatment Bins of Target PPCP Compounds.[1,2,3]

Plant	Plant A	Plant B	Plant C	Plant D	Plant E	Plant F	MBR #01	MBR #02	Bin
Oxybenzone	59 89 92	8 0* 0*	87 83 >96	>96 91 >97	>85 >97 --	98 93 ND	97 95 >99	>85 >97 --	1
Butylbenzyl phthalate	52 >94 >95	0 20 0*	>92 >95 >93	>82 >84 >91	ND 95 --	96 93 >94	97 >97 >96	ND >96 --	1
Methyl-3-phenylpropionate	ND >99.9 ND	ND ND ND	ND ND ND	ND >97 >97	ND ND --	>95 ND ND	ND ND ND	ND ND --	1
Triclosan	0* 64 72	0* 0* 0*	82 60 ND	>89 30 81	ND >96 --	88 86 ND	78 94 >96	ND >96 --	2
Benzophenone	>63 >94 95	62 76 75	91 89 >86	69 85 >98	ND >90 --	>99 ND >79	ND >99 >99	ND 88 --	2
Ethyl-3-phenylpropionate	ND >92 95	>78 0* >89	>14 ND ND	>48 ND ND	ND ND --	>99 ND >63	>76 >73 >74	ND ND --	2
Octylphenol	ND ND ND	ND ND ND	0* ND ND	ND ND ND	ND ND --	>99 ND ND	ND ND ND	ND ND --	2
Galaxolide	ND 0* 0*	21 1 0*	0* 13 0*	0* 32 0	ND 67 --	>86 ND 0*	57 46 36	0* 68 --	3
Musk ketone	0* 0* 10	0* 0* 0*	0* 0* ND	8 0* 0*	0* ND --	85 0* ND	49 36 29	ND ND --	3
TCEP	ND ND ND	0* 3 6	ND 0* ND	ND 0* ND	ND ND --	50 0* ND	ND ND ND	ND ND --	3
DEET	0* 10 21	23 0* 0*	56 26 ND	0* ND ND	ND >84 --	91 >51 >59	ND ND ND	ND >84 --	3
BHA	0 0* ND	ND ND ND	0* 0* ND	ND ND ND	ND ND --	>92 ND ND	>80 >56 >99	ND ND --	3
Triphenylphosphate	ND ND ND	ND 0* 0*	ND ND ND	ND ND ND	ND ND --	57 ND ND	ND ND ND	ND ND --	3

[1] >, indicates that the removal was greater than the percentage value presented because the effluent concentration was below MDL.
[2] ND, not detected in the influent.
[3] 0*, indicates that the effluent concentration was greater than the influent concentration.

Table 3-5 presents a summary distribution of the 20 target PPCPs classified using the occurrence and treatment bins defined in the previous sections. The majority (9 out of 12) of frequently occurring PPCPs were efficiently removed through secondary treatment (i.e., median percent removal >80%). Conversely, Galaxolide, triclosan, and benzophenone were frequently detected but showed medium or low removal by secondary treatment. Other compounds such as octylphenol, TCEP, and triphenylphosphate were not well removed by secondary treatment, but were seldom detected in the WWTP influent at the MDL values of the study. The six compounds included in the shaded area were the most problematic ones, because they were not well removed by secondary treatment but were routinely detected in the influent.

Table 3-5. Distribution of Target PPCPs in Occurrence and Treatment Bins.

Occurrence Bins	Treatment Bins	Compounds	Classification
Bin O1[1] Infrequent	Bin T1 Good removal	Methyl-3-phenylpropionate	Fragrance
	Bin T2 Moderate removal	Octylphenol	Surfactant (detergent)
	Bin T3 Poor removal	TCEP Triphenylphosphate	Fire retardant Fire retardant
Bin O2 Intermediate	Bin T1 Good removal	–	–
	Bin T2 Moderate Removal	Ethyl-3-phenylpropionate	Fragrance
	Bin T3 Poor removal	BHA DEET Musk ketone	Antioxidant Insecticide Fragrance
Bin O3 Frequent	Bin T1 Good removal	Caffeine Ibuprofen Oxybenzone Chloroxylenol Methylparaben Benzyl salicylate 3-Phenylpropionate Butylbenzyl phthalate Octylmethoxycinnamate	Pharmaceutical[1] Pharmaceutical Sunscreen Germicide Antioxidant Sunscreen Fragrance Plasticizer Sunscreen
	Bin T2 Moderate removal	Triclosan Enzophenone	Germicide Sunscreen
	Bin T3 Poor removal	Galaxolide	Fragrance

[1] For compounds in Bin O1 (infrequent occurrence), the treatment classification is limited by insufficient data as they were seldom detected in the influent.

With frequent occurrence and poor removal characteristics, Galaxolide represents one of the most problematic target compounds of the study. Risk assessments of Galaxolide have been conducted in response to the frequent detection of this compound in the environment (The HERA Project, 2004). Although Galaxolide is not readily biodegraded in sewage treatment, it has no demonstrated mutagenic or genotoxic potential and has not shown reproductive changes. It can, however, be mildly toxic to fish, algae, and other invertebrates (International Flavors &

Fragrances, 2003). In addition, there are concerns about bioaccumulation of this chemical in certain aquatic species (i.e., clams and other bivalves) (Oros, 2002). Galaxolide has been shown to be easily photodegraded (Sanchez-Prado et al., 2004). Therefore, although Galaxolide represents a "problematic" compound, its overall environmental risk may be lowered through the application of standard wastewater treatment technologies such as ultraviolet light.

3.2.3 Impact of SRT on the Removal of Soluble PPCPs

The removal efficiencies were plotted against the corresponding SRT for each of the investigated compounds. In general, it was observed that an increase in SRT enhanced the removal of the majority of monitored PPCPs, which agrees with results from other studies (Clara et al., 2005). In this study, a critical SRT was defined as the minimum SRT required for achieving a consistent removal of greater than 80% ($SRT_{80\%}$).

Figure 3-4 presents the removal of Bin T1 compounds in the different treatment facilities. At SRTs higher than 5 days, consistent removal of more than 80% was observed for chloroxylenol, butylbenzyl phthalate, oxybenzone, octylmethoxycinnamate, caffeine, benzyl salicylate, and ibuprofen. Therefore, a critical $SRT_{80\%}$ of about 5 days can be identified for these compounds (i.e., shown as the dash lines in Figure 3-4). Different results were obtained for 3-phenylpropionate, methylparaben, and methyl-3-phenylpropionate, where the effluent concentration was always below MDL at all investigated SRTs. Even though there is some uncertainty about absolute percent removal, particularly for 3-phenylpropionate at lower SRTs, the data strongly suggests that a $SRT_{80\%}$ shorter than 5 days could be appropriate for these three compounds.

The dashed lines in Figure 3-4 that estimate the $SRT_{80\%}$ were located based on observation only. A statistical test was then performed using Student's *t* distribution to verify the hypothesis that the sample data collected below and above the dashed line (critical SRT) appear to be from different underlying populations. Results show that the populations are different with a significance of 99% for cloroxylenol, 98% for oxybenzone, 90% for ibuprofen, and 98% for octylmethoxycinnamate. This test demonstrates that the visual placement of the critical SRT is also statistically valid for these compounds. In the cases of 3-phenylpropionate, methylparaben, and methyl 3-phenylpropionate, no visual estimations of $SRT_{80\%}$ (i.e., dotted lines) were included, because the presence of left-censored data suggests that the value might be even lower than 5 days. In the case of methyl 3-phenylpropionate, the available data is also insufficient to draw a conclusion.

For the compounds in Bin T1, the high removal above a 5-day SRT and poor removal below is similar to the response of nitrifiers. Therefore, it appears that nitrifiers might be doing the removal, or nitrifiers might be closely involved in the co-metabolism of these compounds. Nevertheless, additional data and a full mass balance are needed to test this hypothesis.

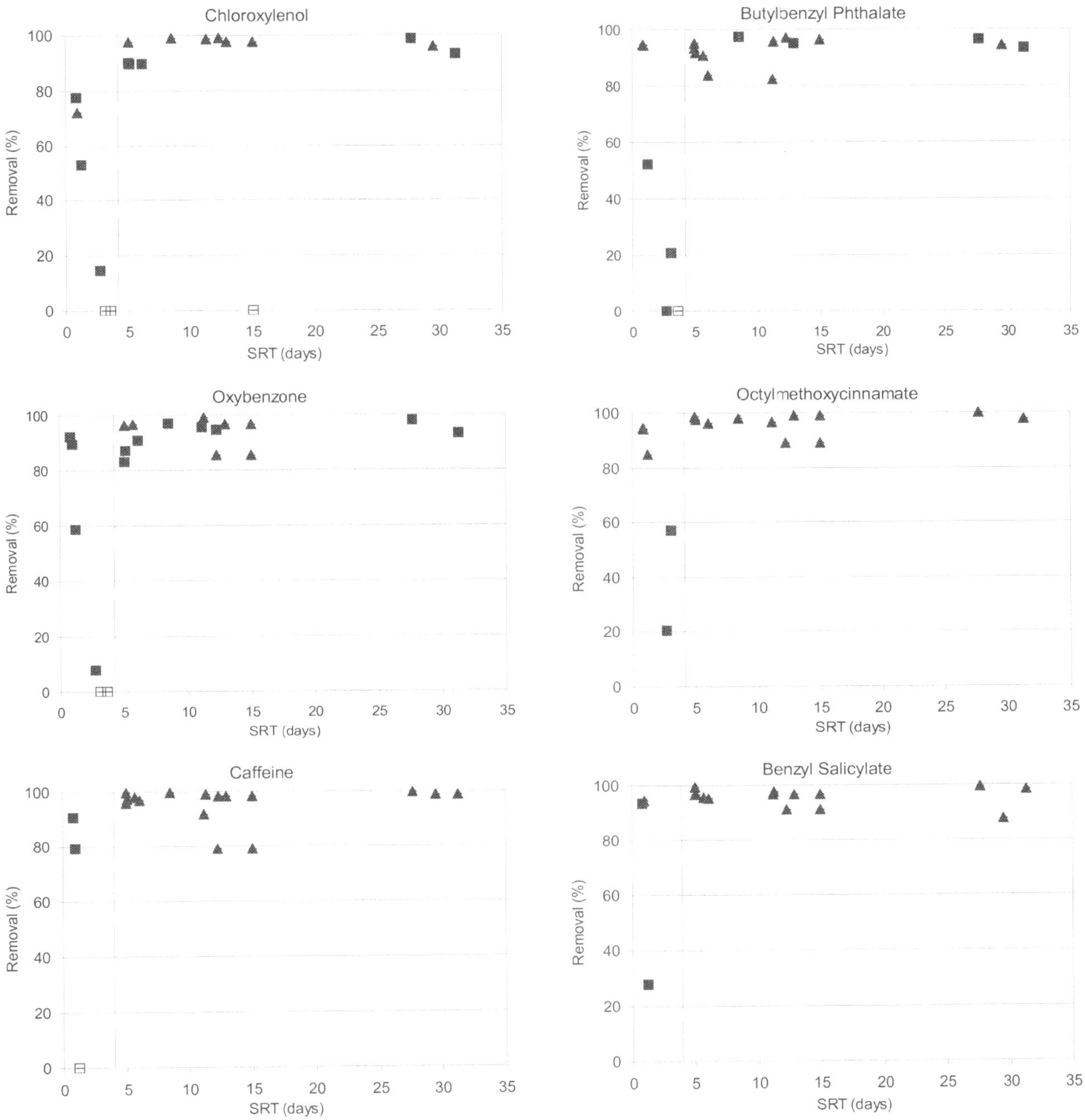

(■--- Actual removal; ▲--- Removal greater than the percentage value presented;
□ --- Effluent concentration greater than influent concentration.)

Figure 3-4. Percent Removal in Relation to SRT for Compounds Susceptible to Activated Sludge Treatment (Bin T1, Good Removal).

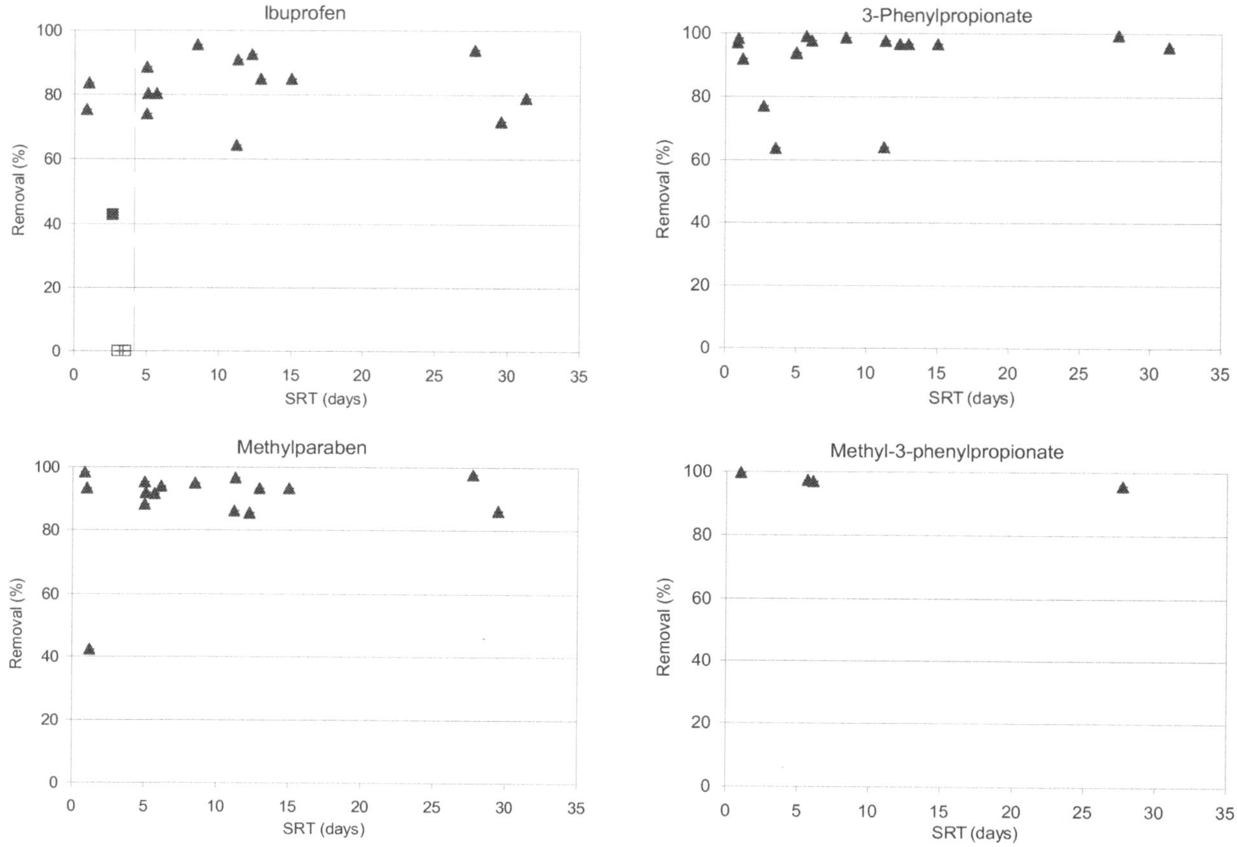

(■--- Actual removal; ▲--- Removal greater than the percentage value presented;
□ --- Effluent concentration greater than influent concentration.)

Figure 3-4 (continued). Percent Removal in Relation to SRT for Compounds Susceptible to Activated Sludge Treatment (Bin T1, Good Removal).

Figure 3-5 shows the dependence of the removal of Bin T2 compounds on SRT. At SRTs higher than 13 days, nearly complete removal of benzophenone was achieved. Similarly, a $SRT_{80\%}$ of 10 days was determined for triclosan. A statistical test for significance followed to justify the visual estimation of $SRT_{80\%}$. Results show that populations separated by the critical SRT line are different with a significance of 99% for triclosan, and 60% for benzophenone. When the treatment facilities were operated at SRTs below the $SRT_{80\%}$, high percent removals were observed at some plants, whereas in others only slight removals were obtained. The $SRT_{80\%}$ for ethyl-3-phenylpropionate is indeterminate because the effluent concentrations of this compound were below MDL at SRTs greater than 5 days. The calculated percent removal underestimated the actual removal because of the left-censored effluent data. This causes the $SRT_{80\%}$ for ethyl-3-phenylpropionate to be indeterminate, but should be greater than 5 days according to the definition of consistent removal greater than 80%. Data were insufficient to identify the $SRT_{80\%}$ for octylphenol removal, which might range between 5 and 28 days.

Figure 3-6 shows the removal dependence on SRT for Bin T3 compounds. PPCPs in this category are poorly removed through biological treatment. Consequently, the $SRT_{80\%}$ values for compounds in this bin are expected to be longer than for compounds in the previous two bins.

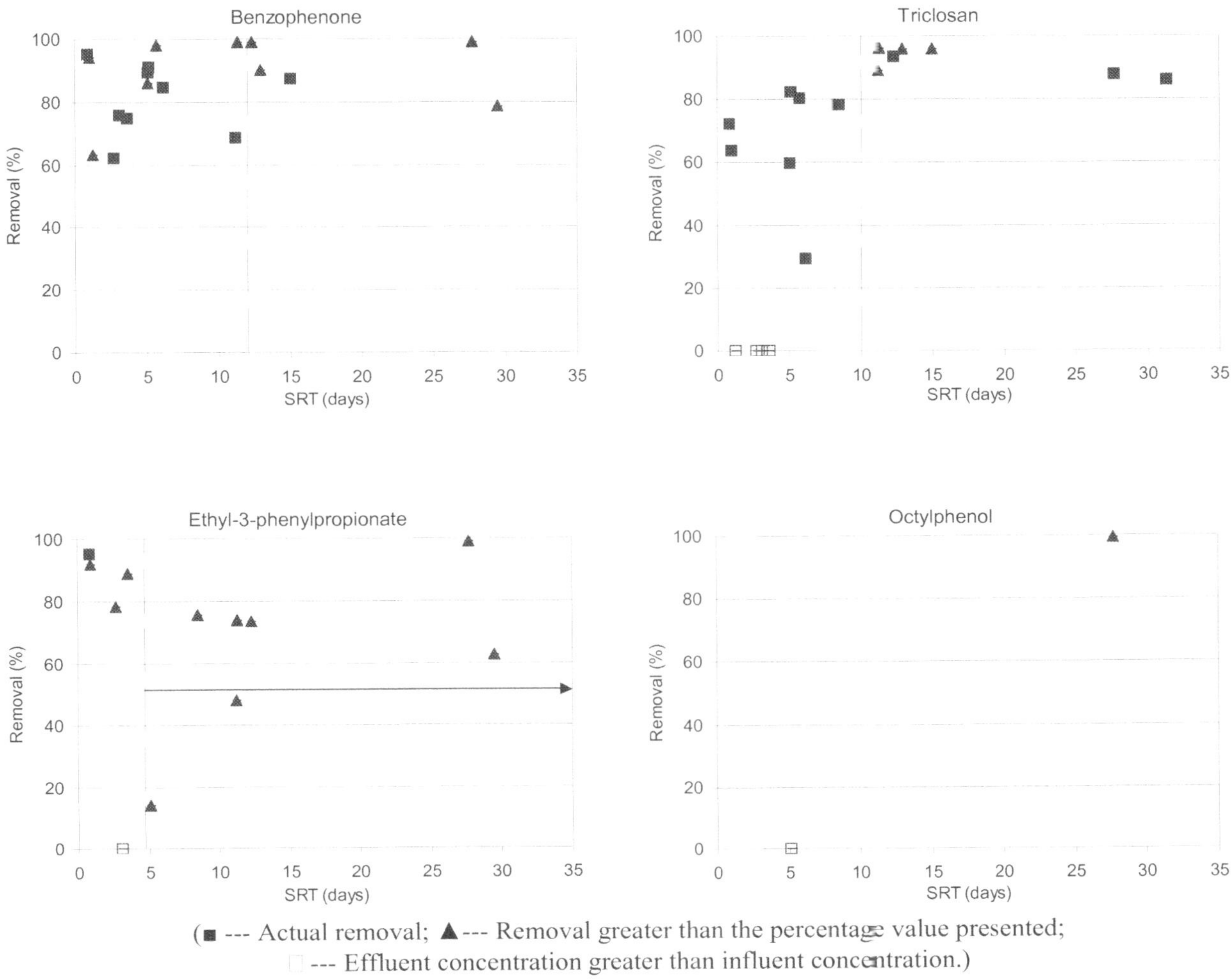

Figure 3-5. Percent Removal in Relation to SRT for Compounds Susceptible to Activated Sludge Treatment (Bin T2, Moderate Removal).

The removal of Galaxolide, musk ketone, and TCEP were enhanced at longer SRTs. However, even at SRT of 30 days, consistently good removal was not observed for these three compounds. $SRT_{80\%}$ values were indeterminate for DEET, BHA, and triphenylphosphate because of insufficient data. It is suggested that the $SRT_{80\%}$ for these three compounds are longer than 13 days, 8 days and 5 days, respectively. The Student's *t* test for DEET using an $SRT_{80\%}$ of 13 days demonstrates a significance of 99%, validating the visual placement of this critical SRT line.

For those compounds that require a much higher SRT for removal, it is possible that there is an acclimation issue or slow kinetic degradation. High SRTs allow the enrichment of slowly growing bacteria and consequently the establishment of a more diverse microbial community with broader degradation capacities. Additional data, however, should be collected and a full mass balance should be performed before specific removal mechanisms can be proposed.

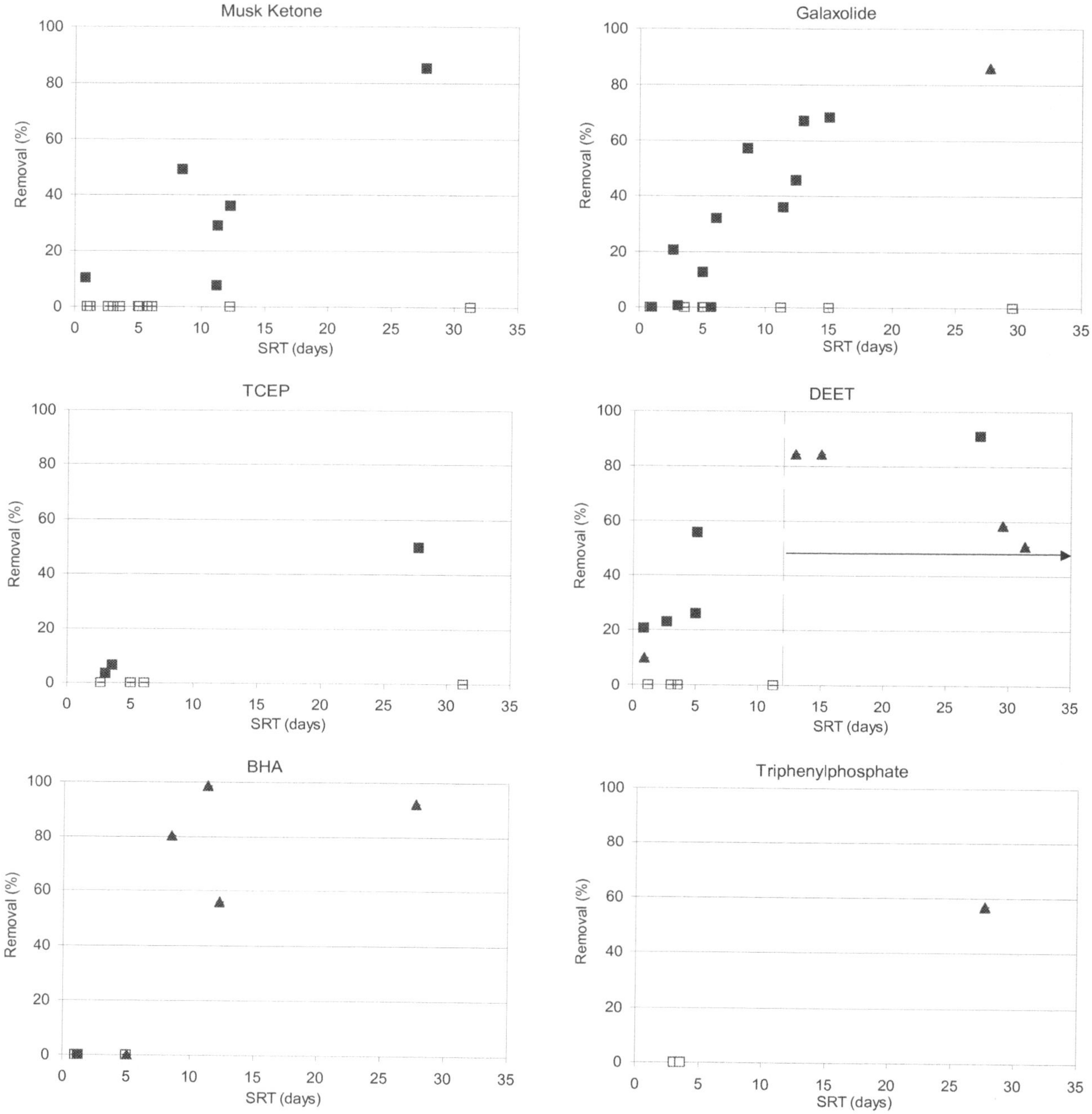

(■ --- Actual removal; ▲--- Removal greater than the percentage value presented;
□ --- Effluent concentration greater than influent concentration.)

Figure 3-6. Percent Removal in Relation to SRT for Compounds Susceptible to Activated Sludge Treatment (Bin T3, Poor Removal).

Table 3-6 summarizes the minimum SRT required for achieving a consistent removal greater than 80% for each analyzed PPCP (i.e., critical SRTs). It should be pointed out that eight target list compounds were listed in the "Method Detection Limits" (Table 2-6) as having matrix background concentrations that were too high, which might lead to artificially low MDL values. For these eight compounds (butylbenzyl phthalate, ethyl-3-phenylpropionate, 3-

phenylpropionate, benzophenone, chloroxylenol, Galaxolide, musk ketone, and TCEP), it is possible that the zero percent removal data points where the effluent concentration is greater than the influent concentration (open squares) have much less confidence than the other data points. Recalculating the $SRT_{80\%}$ for these eight compounds without these data points would lead to the following changes: butylbenzyl phthalate will be lowered from 5 days to 2 days; ethyl-3-phenylpropionate from >5 days to < 5 days; chloroxylenol from 5 days to 4 days; and Galaxolide and musk ketone from >30 days to 15–25 days.

The data suggest that compounds in Bin T1 (i.e., good removal) can be well removed though secondary treatment with a SRT of 5 days or even shorter. Nevertheless, longer SRTs are required to effectively remove Bin T2 compounds. A minimum SRT of at least 10 days is needed

Table 3-6. Critical SRTs to Effectively Remove Target PPCP Compounds.

Bins	Compounds	$SRT_{80\%}$, days[1]
Bin T1 Good removal	hloroxylenol	5[2]
	Butylbenzyl phthalate	5[3]
	Oxybenzone	5
	Octylmethoxycinnamate	5
	Benzyl salicylate	5
	Caffeine	5
	Ibuprofen	5
	3-Phenylpropionate	<5
	Methylparaben	<5
	Methyl-3-phenylpropionate	<5
Bin T2 Moderate removal	Benzophenone	13
	Triclosan	10
	Ethyl-3-phenylpropionate	>5[4]
	Octylphenol	5~28
Bin T3 Poor removal	Musk ketone	>30[5]
	Galaxolide	>30[5]
	TCEP	>30
	DEET	>13
	BHA	>8
	Triphenylphosphate	>5

[1] $SRT_{80\%}$ read from Figures 3-4 to 3-6 is rounded up to the closest integer.
[2] $SRT_{80\%}$ of chloroxylenol will be lowered to 4 days without the open square points.
[3] $SRT_{80\%}$ of butylbenzyl phthalate will be lowered to 2 days without the open square points.
[4] $SRT_{80\%}$ of ethyl-3-phenylpropionate will be lowered to less than 5 days without the open square points.
[5] $SRT_{80\%}$ of Galaxolide and musk ketone will be lowered to 15 to 25 days without the open square points.

to remove 80% of the influent benzophenone and triclosan. For octylphenol and ethyl-3-phenylpropionate in this bin, the $SRT_{80\%}$ should be greater than 5 days, although the exact value is indeterminate because of insufficient data. The compounds in Bin T3 are even more difficult to remove through biological treatment, as there was no evidence of consistent good removal even at SRT of 30 days for musk ketone, Galaxolide, and TCEP. The results also show that

DEET, BHA, and triphenylphosphate cannot be effectively removed at SRTs shorter than 13, 8, and 5 days, respectively. Additional data are needed to identify the $SRT_{80\%}$ for these three compounds. If this data were available, some of these compounds might be classified as Bin T2 (i.e., moderate removal) rather than Bin T3 (i.e., poor removal) compounds.

Some of these results are in agreement with those reported by previous studies. For example, Clara et al. (2005) reported in general the dependence of removal on SRT for certain compounds (not all of them), with more removal being achieved at longer SRTs as a general case. Clara and colleagues also reported a critical SRT of 5 days for buprofen, which agrees with our observations for this compound. On the other hand, Kreuzinger et al. (2004) reported that Galaxolide could be removed at SRTs between 25 to 40 days, which was not consistently observed in our study at the facility operated at an SRT within this range.

3.2.4 Conventional Activated Sludge Processes versus Membrane Bioreactor

When comparing the performance of a conventional activated sludge process to a MBR, it was observed that the latter showed no obvious advantage over the previous one if run at a similar SRT. In this study, two MBRs were run at long SRTs (10–15 days). Consequently, most of the PPCPs were removed to a high degree, especially for the Bin T1 (good removal) and Bin T2 (moderate removal) compounds. Furthermore, for the poorly removed compounds in Bin T3, Table 3-4 indicates that no significant differences in removal efficiency were detected between MBR and conventional activated sludge process when they were operated at similar SRTs. This confirms the expectation that the micro- and ultrafiltration membranes cannot provide additional benefit of removing the PPCP compounds by sieving, as the PPCP molecules are more than 100 times smaller than the pore size of the membranes. In general, these results agreed with those reported by other authors (Clara et al., 2004; Joss et al., 2005)

3.2.5 Effect of Media and Membrane Filtration

For PPCP compounds that were still detected after the secondary treatment, further treatment with media filtration did not provide any additional removal. As presented in Appendix G, the removal efficiencies of these compounds through the media filters in Plant D and Plant F were very inconsistent (and were even negative in some cases).

When using RO treatment instead of media filtration, all the analyzed PPCPs were not detected in the RO permeate at MDL values (Appendix G). These results were consistent and fully support RO is as an effective barrier for preventing the passage of PPCP compounds present in secondary wastewater effluent.

3.3 Qualitative Observations

In addition to the key findings presented in the previous section, an examination of the occurrence and treatment data resulted in some interesting qualitative observations. Insufficient data were collected to consider these findings as valid conclusions, although additional studies might be warranted. These qualitative observations are discussed in the following sections.

3.3.1 Seasonal Variations of PPCP Occurrence

Plants were sampled in summer and winter months to investigate seasonal variations in PPCP occurrence. Sampling events were conducted from August through October (summer samples) and from November through March (winter samples). Constraints (i.e., rain events, plant operation issues) prevented an equal distribution of seasonal samples, and only one-fifth of the samples were collected during summer months. Although the data might be insufficient to support any definitive conclusion regarding seasonal effects, it revealed some general trends that may be of interest.

The seasonal variation of PPCPs in secondary influent was investigated separately for plants with and without primary treatment. Figure 3-7 shows the occurrence of 3-phenylpropionate on different sampling dates in *plants without primary treatment* (Plant E and Plant F). It was observed that the concentration of 3-phenylpropionate in Plant F was higher in summer than in winter, and the same trend was observed for all the 20 monitored compounds (i.e., as presented in Appendix H). However, Figure 3-7 also shows a great variation of the 3-phenylpropionate concentration in Plant E, where the two sampling events occurred just two days apart. This variation of about 1 log is smaller than the observed seasonal variation of around 2 logs. To further evaluate the suggested trend, additional data would need to be collected.

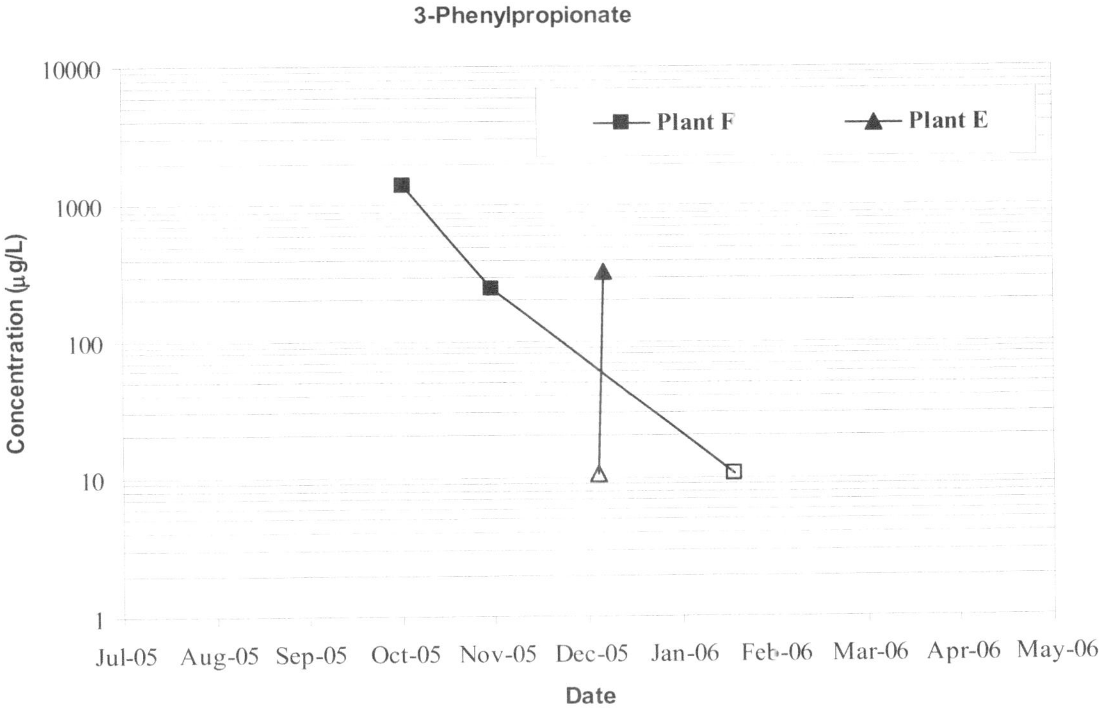

(Open symbols indicate concentration below MDL; lines only to aid the observation)

Figure 3-7. Occurrence of 3-phenylpropionate in Plants without Primary Treatment.

A different trend was observed when analyzing PPCP concentrations in *plants with primary treatment*. For example, 3-phenylpropionate was detected at higher concentrations in winter (shown in Figure 3-8). The same trend (i.e., influent concentration higher in winter), was observed for other compounds such as methylparaben, caffeine, Galaxolide, oxybenzone, etc. (i.e., as presented in Appendix I). However, noticeable variations in PPCP concentration were observed in sampling events that were conducted close together, as the case with Plant B and MBR #01, as presented in Figure 3-8. The observed concentrations during seasonal sampling events showed a variation of approximately 2 logs, although this variation was just 0.5 log when the sampling events were close together. Collection and evaluation of additional data would be needed to substantiate the suggested seasonal variation.

Despite the data limitations, there was evidence of seasonal variation of PPCP concentration in secondary influent with the influent concentrations decreasing in cooler months for plants without primary treatment and the reverse trend of increasing for plants with primary treatment. Additional data is required to fully substantiate this observation.

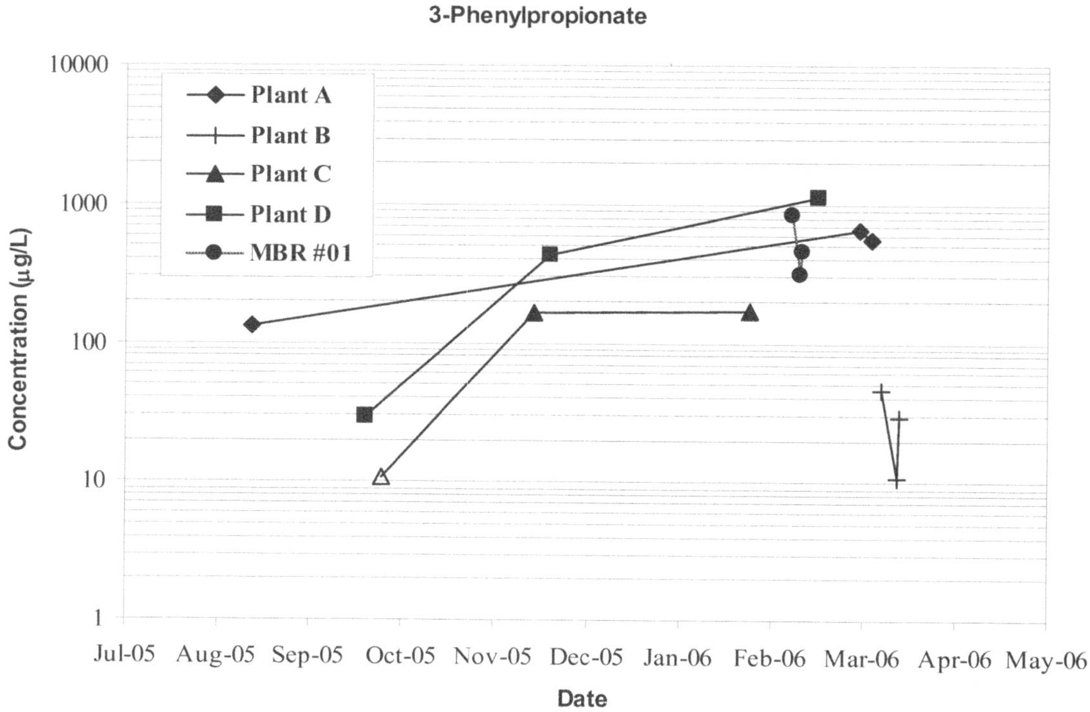

(Open symbols indicate concentration below MDL; lines only to aid the observation)

Figure 3-8. Occurrence of 3-phenylpropionate in Plants with Primary Treatment.

3.3.2 Impact of HRT on the Removal of Soluble PPCPs

HRT and SRT are both important design parameters for WWTPs, and HRT usually correlates well with SRT (shown in Figure 3-9). However, when a plant is operated below design capacity (i.e., as it was the case of Plant E shown in Figure 3-9), HRT could divert from this linear relation with an abnormally high value. Conversely, MBRs allow achieving high SRTs within a compact reactor, resulting in smaller HRTs as compared to conventional treatment systems.

The impact of HRT on PPCP removal was investigated by analyzing plants of similar SRT but different HRTs. In this study, Plant D, Plant E, MBR #01, and MBR #02 were operated under similar SRTs (11–15 days) but distinct HRTs (5–30 h). The removal efficiency of PPCP compounds was compared among these four plants to investigate the possible benefit of operating at longer HRTs (i.e., longer contact time between wastewater and biomass).

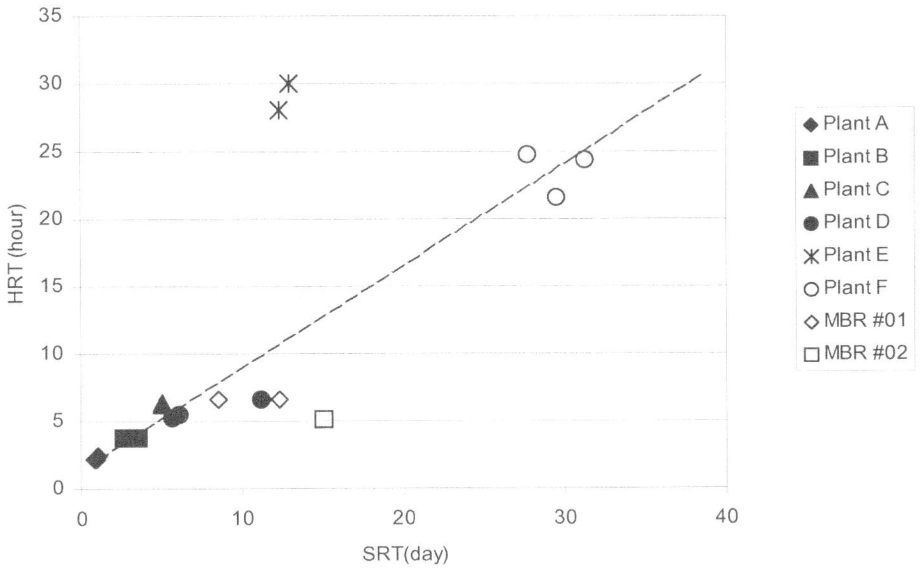

Figure 3-9. Correlation of HRT and SRT in Sampled Facilities.

The results do not indicate that HRT has much influence on the percent removal. Actually, most of the compounds were completely removed at the investigated conditions, considering that the applied SRT was longer than 10 days. For the poorly removed compounds in Bin 3 (i.e., Galaxolide and musk ketone shown in Figure 3-10), there was no apparent impact of HRT on the removal efficiency. However, general conclusions cannot be made without additional data.

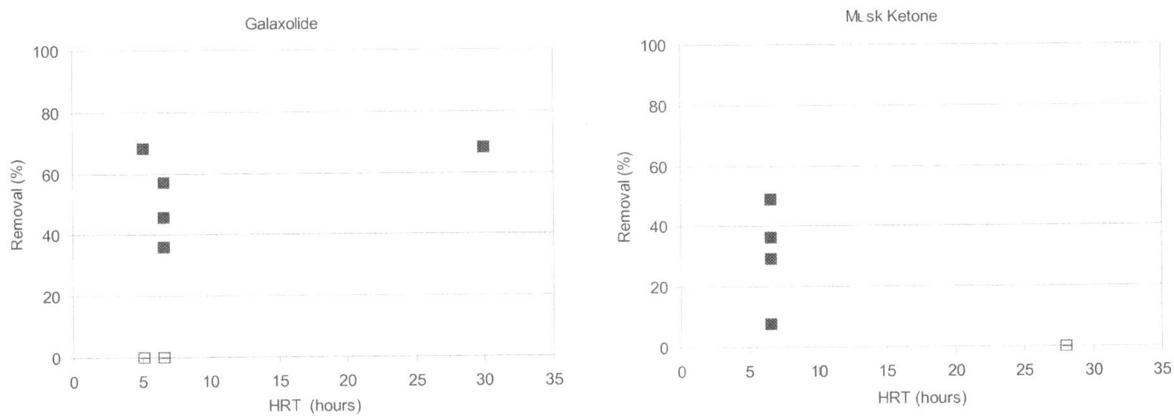

Figure 3-10. Percent Removal in Relation to HRT for Galaxolide and Musk Ketone.

Table 3-7. K_{ow} of PPCP Compounds.

Bins	Range of Log K_{ow}[1,2,3,4]	$SRT_{80\%}$
Good removal	-0.07 ~ 4.31	>5
Moderate removal	2.73 ~ 4.76	>10
Poor removal	1.44 ~ 5.9	>30

[1] Hansch et al., 1995; [2]. Balk & Ford, 1999; [3]. Tas et al., 1997; [4]. Ahel & Giger, 1993.

3.3.3 Other Factors Potentially Impacting the Removal of PPCPs

Pure Oxygen

As shown in Table 3-4, Plant A (high-purity oxygen process) generally performed better than Plant B (conventional activated sludge) in terms of PPCP removal, even though Plant A was operated at shorter SRTs (i.e., 1 day vs. 3 days). However, the available data were insufficient to establish a definitive conclusion from this observation.

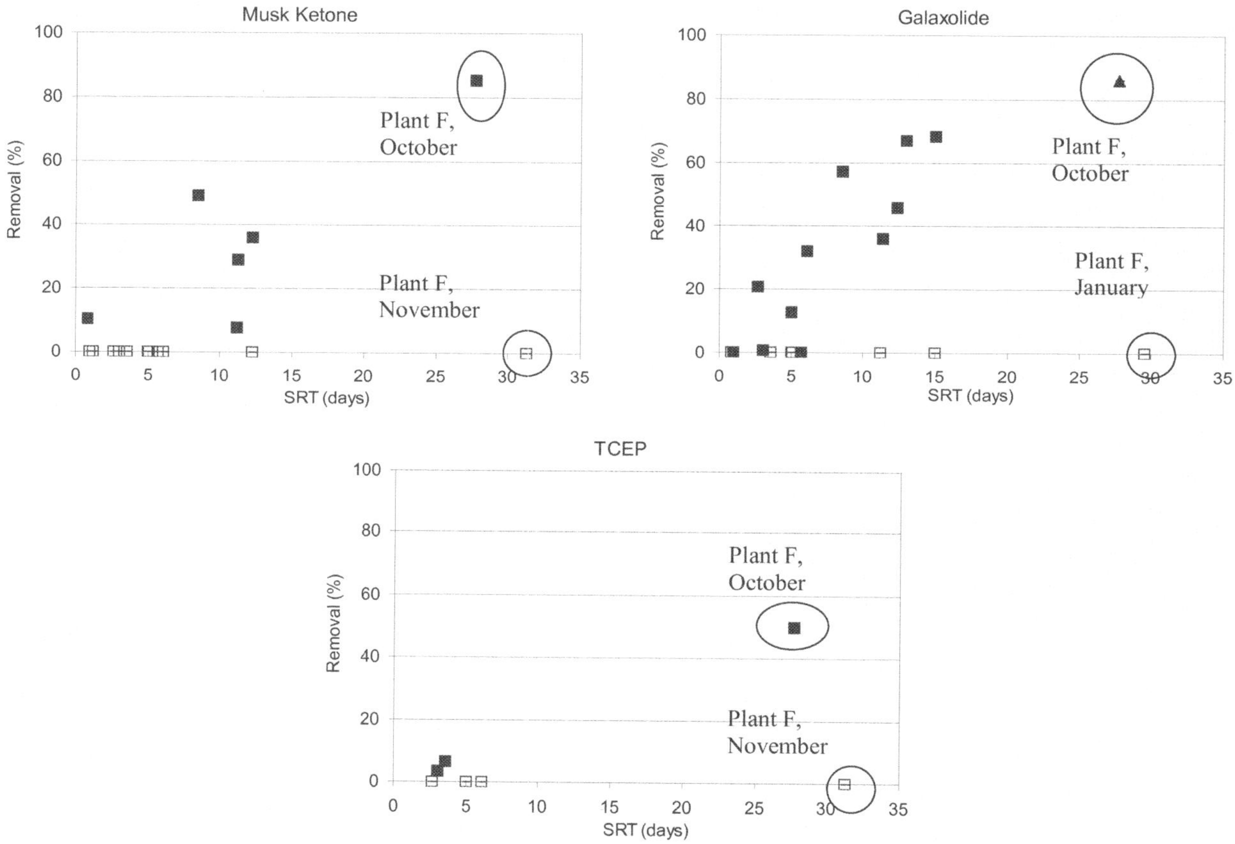

(■ --- Actual removal; ▲--- Removal greater than the percentage value presented;
□ --- Effluent concentration greater than influent concentration.)

Figure 3-11. Seasonal Variation in the Removal of Musk Ketone, Galaxolide, and TCEP in Plant F.

Octanol-Water Partition Coefficient (K_{ow})

The results from this study suggest that there is no apparent correlation between the removal of soluble PPCPs and the K_{ow} of the analyzed compounds. K_{ow} is an indicator of the tendency of a chemical to partition itself into an organic phase (hydrophobicity) or an aqueous phase (hydrophilicity). Compounds with a log K_{ow} below 1.0 are very hydrophilic, whereas a value of 4.0 or larger indicates that the compound is very hydrophobic. As shown in Table 3-7, a wide range of K_{ow} values was identified for the compounds in the same removal bins. This observation was expected, because the analytical methods for PPCP detection and quantification were solely focused on the aqueous phase, whereas K_{ow} is generally considered as a parameter indicating the adsorption of PPCP into the biosolids. To further investigate the PPCP removal through secondary treatment in relation to K_{ow}, the PPCP concentrations in the solid phase would need to be determined and a mass balance analysis conducted.

Seasonal Variations

Plants A, C, D, and F were sampled both in winter and summer months, and the seasonal variation of PPCP removal in these four facilities was investigated. When reviewing the percent removal data from Table 3-4, no significant differences in removal efficiency were observed between warmer and cooler seasons for Plants A, C, and D. A seasonal effect would probably not be evident in these plants because their operating SRT is still too low to achieve much removal of these compounds. However, significant variations of musk ketone, Galaxolide, and TCEP removal were observed in Plant F at SRTs of around 30 days, as shown in Figure 3-11. The wide variation in removal observed for these compounds at an SRT of 30 days could not be attributed to any apparent differences in plant operations. The good removals were all attained during an October 5 sampling event and the poor removals were all attained during November 12 or January 9 sampling events. The plant daily effluent discharge temperatures for these dates, however, do not support a seasonal influence on removal because recorded temperatures were 25.8°C on October 5, 24.4°C on November 12, and 18.9°C on January 9. A mass balance study is needed to assess whether longer SRT values can result in desorption from the sludge.

CHAPTER 4.0

CONCLUSIONS AND RECOMMENDATIONS

This study was based on the collection of representative samples of influent and effluent wastewater samples from a group of WWTPs (and pilot unit MBRs) that exhibit a wide range of SRTs. The collected samples were analyzed to evaluate both the occurrence of a target list of PPCP compounds that are routinely detected in municipal wastewater effluent, and the degree of removal of these chemicals through conventional secondary treatment. Because the analytical method included a filtration step (i.e., using 0.45-μm filters), the reported results were solely based on the analysis of the aqueous phase. Conclusions regarding occurrence and treatment constitute the core of this section, which also includes a group of supplemental conclusions based on qualitative observations.

4.1 Occurrence

A group of samples were collected from the influent to secondary treatment from a selected group of WWTPs and analyzed to determine the concentration of 20 specific PPCPs. Overall, the detected concentrations of these chemicals were between 0.013 and 1380 μg/L. The compounds were classified into three occurrence bins based on the percentage of observations that exhibited a detectable concentration. Compounds selected within each occurrence bin and the corresponding selection criteria are presented in Table 4-1.

Table 4-1. Classification of Analyzed PPCPs in Occurrence Bins.

Occurrence Bins	Criteria	Compound	Classification
Bin O1 Infrequent	Detected < 25%	TCEP Triphenylphosphate Octylphenol Methyl-3-phenylpropionate	Flame retardant Flame retardant Surfactant (detergent) Fragrance
Bin O2 Intermediate	25% <Detected < 75%	Musk ketone Ethyl-3-phenylpropionate BHA DEET	Fragrance Fragrance Antioxidant Insecticide
Bin O3 Frequent	Detected > 75%	Galaxolide 3-Phenylpropionate Cloroxylenol Triclosan Caffeine Ibuprofen Benzophenone Benzyl salicylate Octylmethoxycinnamate Oxybenzone Methylparaben Butylbenzyl phthalate	Fragrance Fragrance Germicide Germicide Pharmaceutical Pharmaceutical Sunscreen Sunscreen Sunscreen Sunscreen Antioxidant Plasticizer

As indicated in this table, the majority of analyzed PPCPs (60%) were classified as part of Bin O3 (i.e., frequent occurrence); the rest were evenly distributed between the other two bins. Flame retardants were among the compounds with the lowest frequency of detection; sunscreens, fragrances, and pharmaceuticals were frequently detected in the influent to secondary treatment.

4.2 Treatment

The concentration of each of the target PPCPs was also analyzed in the effluent of the WWTPs, to compare it to the corresponding influent concentration and determine the respective degree of removal. In addition, plant operational data corresponding to the sampling dates and monthly averages were obtained. This information was used to establish potential relationships between the degree of removal of a given compound and the SRT of the WWTP.

The compounds were then classified into three treatment bins based on their percent removal. As mentioned before, these results were exclusively based on aqueous phase analysis. Compounds included within each of these bins and the corresponding selection criteria are presented in Table 4-2.

Fifty-five percent of the analyzed compounds exhibited good removal; 20% and 25% of the compounds, respectively, exhibited moderate and poor removal. Flame retardants and some antioxidants, insecticides, and fragrances were classified among the compounds that showed poor removal.

Table 4-2. Classification of Analyzed PPCPs in Treatment Bins.

Occurrence Bins	Criteria	Compound	Classification
Bin T1 Good removal	MPR > 80%	Methyl-3-phenylpropionate 3-Phenylpropionate Chloroxylenol Caffeine Ibuprofen Butylbenzyl phthalate Benzyl salicylate Octylmethoxycinnamate Oxybenzone Methylparaben	Fragrance Fragrance Germicide Pharmaceutical Pharmaceutical Plasticizer Sunscreen Sunscreen Sunscreen Antioxidant
Bin T2 Moderate removal	50% < MPR < 80%	Benzophenone Octylphenol Triclosan Ethyl-3-phenylpropionate	Sunscreen Surfactant (detergent) Germicide Fragrance
Bin T3 Poor removal	MPR < 50%	TCEP Triphenylphosphate BHA DEET Musk ketone Galaxolide	Flame retardant Flame retardant Antioxidant Insecticide Fragrance Fragrance

MPR,= median of the percent removal.

When comparing treatment and occurrence results as shown in Table 4-3, it was found that 9 out of the 12 frequently detected compounds also showed good removal, which suggests that secondary treatment can be considered overall as an effective first barrier to the passage of the investigated PPCPs. Combining the results of the occurrence and treatment matrices also allowed identifying a group of compounds that show medium and high occurrence in wastewater effluent, and that at the same time exhibit poor and moderate removal by conventional secondary treatment. These compounds are highlighted in boldfaced type in Table 4-3, and include fragrances, sunscreens, insecticides, antioxidants, and germicides. Although no health hazards have been associated with these compounds, their frequent detection in wastewater effluent and the medium to poor ability of secondary treatment to remove them may be a future environmental concern if they are shown to contaminate food supplies or damage ecosystems. Despite exhibiting poor removal by secondary treatment, the analyzed flame retardants (TCEP and triphenylphosphate) were not included in this group, because they were very infrequently detected at concentrations above the MDL level.

Table 4-3. PPCPs of Concern Based on Occurrence and Treatment Bins.

Treatment / Occurrence	Bin T1 Good Removal	Bin T2 Moderate Removal	Bin T3 Poor Removal
Bin O1[1] Infrequent	Methyl-3-phenylpropionate	Octylphenol	TCEP Triphenylphosphate
Bin O2 Intermediate		Ethyl-3-phenylpropionate	**BHA** **DEET** **Musk ketone**
Bin O3 Frequent	Caffeine Ibuprofen Oxybenzone Chloroxylenol methylparaben Benzyl salicylate 3-Phenylpropionate Butylbenzyl phthalate Octylmethoxycinnamate	**Triclosan** **Benzophenone**	**Galaxolide**

[1] For compounds in Bin O1 (infrequent occurrence), treatment classification is limited by insufficient data as they were seldom detected in the influent

4.2.1 Dependency of PPCP Removal on SRT

This analysis also revealed specific trends that relate the degree of PPCPs removal through secondary treatment to the specific SRT of the conventional biological system. The analyzed PPCPs appear to classify into three different groups in terms of the critical SRT, which in this study is defined as the minimum SRT required for achieving a minimum percent removal of 80% ($SRT_{80\%}$).

- *$SRT_{80\%} \leq 5$ days*
  Includes all compounds from Bin T1. Methyl 3-propionate and 3-phenylpropionate may have a very low $SRT_{80\%}$.

- *$SRT_{80\%}$ > 5 days and <30 days*
  Includes all compounds from Bin T2, with the exception of ethyl 3-phenylpropionate and octylphenol, in which cases insufficient data were available to discern a required minimum SRT (i.e., the data suggests it occurs somewhere between 5 and 30 days)

- *$SRT_{80\%}$ > 30 days*
  Includes all compounds from Bin T3, where removal shows a strong dependence on SRT. However, these compounds sometimes show very poor removal even at SRTs of 30 days or higher. Insufficient information is available to completely characterize BHA and triphenylphosphate as part of this group.

These results suggest a dependence of PPCP removal on SRT that is somewhat compound specific. In general, larger percent removals were observed at longer SRTs, although there were a handful of compounds that showed poor removal even at SRTs longer than 30 days. The conclusions from this section agree with other studies that establish that the correlation between PPCP removal and SRT depends on the analyzed compound (Joss et al., 2005), the general case of higher percent of PPCP removal observed at longer SRTs (with certain exceptions) (Kreuzinger et al., 2004) and evidence that that significant removal of PPCPs may be achieved in WWTPs operating at SRTs longer than 10 days (Clara et al., 2005; Joss et al., 2004; Kreuzinger, et al., 2004).

4.2.2 Other Conclusions

Additional experimental data collected during this study was sufficient to support the following supplemental conclusions:

- *Performance of MBRs vs. Conventional Secondary Plants*
  No significant differences were observed when comparing the performance of pilot-scale MBR units and conventional secondary plants, when they operated at SRTs longer than 10 days. This finding has been also reported by other studies (Clara et al., 2004).

- *Media Filtration*
  When PPCP compounds were present after secondary treatment, further treatment via media filtration did not offer any additional benefit. This was an expected outcome, as the analytical method only takes into account the dissolved chemicals present in the aqueous phase, which should not be removed by conventional media filters.

- *RO Treatment*
  When PPCP compounds present in secondary effluent were subject to RO treatment, the analyzed PPCPs were effectively removed to concentrations below the MDL. This occurred even for those compounds not consistently removed at an SRT of 30 days (i.e., Galaxolide). This conclusion is in agreement with reports from other investigators (Snyder, 2003b; Everest et al., 2003).

Finally, additional information obtained from this study suggests other potential trends but the available data were not sufficient to fully support conclusions without further studies. Some of these trends included the following:

- *Influence of HRT*
  When investigating the effect of HRT in PPCP removal in plants operating at similar SRT, the available data suggest that HRT does not have much impact on the degree of removal of the investigated PPCP compounds.

- *Pure Oxygen Systems vs. Conventional Aeration Systems*
  When comparing the PPCP removal performance of WWTPs operating below the suggested critical SRT of 5 days, the pure oxygen system performed better than the conventional aeration system even though the SRT of the pure oxygen plant was slightly shorter than the conventional aeration plant.

4.3 Future Work

A complete understanding of the implications of PPCP release to the environment will require extensive data, which are presently being compiled by the industry. The wastewater treatment facility occurrence and treatment data provided by this study contributes valuable information toward this complex issue, but additional efforts are still required. A suggested guideline for future work includes expanding the occurrence and treatment database by including information from additional WWTPs, increasing the target list of PPCPs, and reducing the detection limits of the analytical methods used for quantifying these compounds. These efforts will remain incomplete, however, until reliable information on the potential health effects and ecological risks associated with the presence of PPCPs in the environment becomes available. This information will make possible a better assessment of the environmental risks associated with these chemicals. The public and the wastewater treatment community will benefit from this information because they will be able to develop more efficient treatment strategies and stewardship programs to minimize the release of PPCPs potentially harmful to the environment.

APPENDIX A

SAMPLING EQUIPMENT SPECIFICATIONS

Glacier® Composite Sampler
Teledyne Technologies Incorporated
12333 West Olympic Boulevard - Los Angeles, CA 90064

The Glacier® Sampler combines the small size, lightweight, and mobility of a portable unit with an exclusive new active temperature control system. Its revolutionary design gives you the best of all worlds: easy transport, quick setup, and accurate sample preservation—without reliance on ice or utility power! An optional Mobility Kit with aluminum frame and pneumatic tires lets you wheel Glacier even over rough terrain to remote sampling sites.

In the field, Glacier delivers 48 hours or more of refrigeration from a 12-volt deep cycle battery. Its power-saving cooling system can be set to standby until the first sample is drawn. Glacier can wait patiently for days or weeks to collect event-triggered samples, and then preserve them until a convenient pickup time. For even greater versatility, Glacier also plugs into a 120 or 240vAC outlet to run on AC line voltage.

Unique Temperature Control System

Glacier's new patent-pending temperature control system continuously measures the actual sample temperature, and precisely regulates cooling to ensure accurate sample preservation.

Standard Features

- Dependable, easy-to-program controller uses field-proven technology to ensure the reliability, ease of operation, and durability expected from ISCO.
- Runs on either external 12vDC or 120/240vAC mains power.
- Recall and run a stored program with two keystrokes.
- NEMA 4x, 6 (IP67) controller enclosure is watertight and dustproof.
- Actual measured sample temperature is stored by controller. Minimum, maximum, and average temperatures are reported on LCD and in downloadable file using Samplink software.
- Patented LD90 noncontacting Liquid Presence Detector and pump revolution counting system ensure accurate sample volumes.
- Collect composite samples in 10- or 20-L bottles—or eliminate bottle washing with ProPak™ single-use liners.

Applications

- Stormwater sampling
- Biomonitoring—where accurate sample preservation is needed
- Influent and effluent sampling—Glacier can be easily moved in-plant to collect monthly or weekly at multiple sampling points
- Anywhere you need sampling mobility plus reliable, ice-free cooling

Options and Accessories

- Mobility option—includes aluminum frame with handle and pneumatic tires
- Single-bottle sampling configurations- 1 x 10-L plastic or glass; 1 x 20 L plastic, or 9-L ProPak single use liner system
- Suction lines and strainers

Specifications

- Size: 25 x 15 x 24 inches (63 x 38 x 60 cm)
- Weight: Dry, less battery - 53 lbs (23.5 kg)
- Bottle configurations: 1 – 10 L PE or Glass; 1–20 L PE; 1–9 L ProPak composite sample container with disposable liners
- Power Requirements: 12vDC (Supplied by external battery), or 120/240vAC power with included line cord.
- Cooling: CFC-free refrigeration system maintains sample temperature at 3°C ±1°C at ambient temperature of 0-40°C.

Pump

- Intake purge: Purge before and after each sample.
- Tubing life indicator: Provides a message to change pump tubing.
- Intake suction tubing: Length 3–99 ft. (1–30 m), Vinyl or Teflon® lined, ID 0.375 in. or 0.250 in
- Pump tubing life: 2,000 samples
- Typical repeatability: ±10 ml
- Maximum lift: 26 ft. (7.9 m)
- Typical line velocity @ Head height: 2.9 ft./s @ 3 ft; 2.5 ft./s @ 10 ft; 1.9 ft./s @ 15 ft. (0.88 m/s @ 0.9 m; 0.76 m/s @ 3.1 m; 0.58 m/s @ 4.6 m)
- Sample presence detector: Non-wetted, non-conductive sensor detects when liquid sample reaches the pump to automatically compensate for changes in head heights.

Controller

- Weight: 8 lbs. (3.6 kg)
- Size: 10 x 12.5 x 10 inches (26 x 32.x 25 cm)
- Operational temperature: 32- 120°F (0- 49°C)
- Enclosure rating: NEMA 4X, 6 (IP67)
- Program memory: Non-volatile ROM
- Flow meter signal iNDut: 5- 15vDC pulse or 25 millisecond isolated contact closure for flow-paced sampling.
- External trigger: Initiates sample collection and cooling from contact closure signal
- Number of samples to shutoff: 999 samples maximum with 5-gallon (19-L) bottle.
- Clock Accuracy: 1 min per month, typical, for real time clock

Software

- Sample frequency selection: 1 min to 9,999 min, in 1 min increments. 1 to 9,999 flow pulses
- Sampling modes: Time or flow, event. (Flow mode is controlled by external flow meter pulses.)

- Programmable sample volumes: 10-9,990 ml in 1 ml increments
- Program storage: One program running, one stored
- Controller diagnostics: Tests for RAM, ROM, pump, and display

APPENDIX B

SAMPLING PROTOCOL FOR 24-HOUR TIME-COMPOSITE SAMPLES

Background

The objective of the sampling protocol was providing a snapshot of the water quality conditions during a 24-h period at the selected WWTPs. Because of the extremely low concentration of the investigated analytes (i.e., in parts per billion or parts per trillion) and the high sensitivity of the analytical equipment, minimization of all possible sources of sample contamination was critical. Examples of cross-contamination include but are not limited to products such as deodorant, soap particles, and shampoo residue carried by the sampling operator. The sensitivity of the reading is so high that even residuals of mouthwash and toothpaste carried by the breath of the operator may be sources of sample contamination. The following protocol is based on a previous procedure developed by Snyder, 2005.

Materials

The following materials were compiled and organized before starting each of the 24-h sampling events.

- Powder-free Clean room Nitrile gloves, two dozen
- Plastic 1-L graduated cylinder
- Acetone, 2 gallons
- Hexane, 2 gallons
- Non-reactive glass container for storage of used cleaning chemicals
- Dark Amber 1-L sample bottles for PPCP testing (two per sample point and two additional for collecting blank samples).
- Plastic 5-gallon cubitainer for additional water quality testing (i.e., BOD, COD, TSS, and VSS testing, one per sampling point)
- One 36-quart ice chest per each eight dark amber sample bottles collected
- One 50-quart ice chest per each two cubitainers collected
- Bubble wrap for sample shipment
- Glacier 24-h sampling equipment chest
- One 2.5-gal sample jar per sampling site
- High-quality, nonreactive tubing. 3/8" (quantity varies depending on basin depth
- Nozzle strainer for sampling tubing, one per sampling point
- Duct tape
- Frozen Blue Ice packs ("Blue Ice"; sufficient for shipping all samples)
- One 20-gallon plastic container for deionized water

Procedure

Each 24-h sampling event required a four-step process, which included the following steps:

1. Set-up of equipment
2. 24-h sampling process (including collection of splits and blank samples)
3. Sample packaging and shipment
4. Equipment demobilization and clean up

Set-up of Equipment

The Set-up and Assembly procedure itself is organized into four subtasks: (1) acquisition of materials, (2) selection of sampling site, (3) cleaning of sampling materials, and (4) assembly of equipment.

Subtask 1 (Acquisition of materials). Involves collecting all materials required for the sampling process as described in the Materials section.

Subtask 2 (Selection of sampling site). After all required material is available and transported to the facility, the sampling point must be selected in coordination with operators and senior personnel from the facility. Variable information need to be obtained included, but is not limited to: (a) possible sampling points; (b) variable height of sampling basin over the 24-h period for each possible site; (c) velocity of water at each sampling point. Such considerations are important, as complications may arise during the sampling period. For example, the required pump head may change depending on the variable basin depths. Also, sampling points with high peak flowrates should be avoided, as the moving water may remove the influent tubing from the testing apparatus, and in some cases even drag the sampling equipment with it.

Subtask 3 (Cleaning of sampling materials). The Glacier sampling chest and all other sampling materials must be subject to chemical cleaning with acetone and hexane before starting the sampling process (approval of the facility staff is required in advance). The sampling operator is require to clean the following materials: (a) 2.5-gallon sample jar, (b) influent tubing, (c) peristaltic pump tubing, (d) peristaltic to sample jar tubing, and (e) influent strainer nozzle. Handling of all pieces of equipment should be performed while wearing powder-free Nitrile gloves, which must be replaced by new ones at the end of each cleaning step. Each piece of equipment is cleaned first with acetone followed by hexane using the procedure outlined below.

a) For the 2.5-gallon sample jars

Pour approximately 500 mL into the jar. Place the associated jar screw top onto the 2.5-gallon container and shake and swirl the cleaning chemical for 1–3 min. After thoroughly coating the interior of the jar with the cleaning chemical, pour the contents into the seal-able glass chemical container predetermined for wasted chemicals. Repeat this process a minimum of three times for each jar. After completing acetone cleaning, perform the same process using hexane.

b) For all three pieces of sampling tubing

Fill the nonreactive tubing with acetone and allow it to sit for two to three min. Dump all rinsed acetone into a chemical waste container. Refill tubing with acetone and position both ends vertically with duct tape so that the entire tubing interior is submerged in the chemical. Let tubing sit for 30 min, and dump acetone into chemical waste container. Repeat three times. Repeat this process using hexane. Note that it is required to change gloves between chemical cleans and that all used chemical must be properly placed into a non-reactive glass container. Used hexane may be mixed with used acetone in the chemical storage container. This cleaning process usually takes a minimum of four hours to complete.

c) For the metal strainer nozzle
Place the nozzle into a glass 1-L graduated cylinder, and fill it with acetone. Allow nozzle to sit in acetone for 30 min, after which time the chemical should be wasted and refilled. Perform this process three times and repeat with hexane.

d) For the internal walls of the Glacier sample apparatus
Clean the interior walls of the chest with acetone using industrial grade paper towels. Repeat the process with hexane. These surfaces should not be in contact with the sample, but should still be properly maintained.

Subtask 4 (Assembling of equipment). After performing the cleaning procedures, the operator may then assemble the equipment. It is important to note that the nozzle to influent tubing connection and the influent tubing to peristaltic pump tubing connection must be made using pipe clamps. Use a screwdriver to securely tighten the tubing connection to avoid the loss of equipment into the sample basins.

24-h Sampling Process

After the equipment has been properly cleaned, assembled, and set up at the sampling site, the 24-h sampling period may start. To perform this process, follow the steps listed below using the Glacier sampling equipment control panel:

- Use the Glacier command screen to turn on equipment
- Select "Program"
- Select "Time Based"
- Select "60 min between". Indicates time between samples
- Select "9600 ml". Indicates total size of sample collected
- Select "Take 24 Samples". Indicates total number of samples
- Select "400 mL". Indicates size of individual sample
- Select "No Delay to Start". Unless the operator wishes to begin sampling a preset amount of hours later.

At this point, the operator will be prompted to calibrate the pumping system.

- Select tubing size. Should be 3/8" x length of tubing in feet
- Select "Yes" to the question "Check Vol. Sample?"

After selecting the "yes" option, the Glacier apparatus begins the sampling procedure for calibration purposes. The Glacier sampling system takes samples in the following manner: First, the pump purges the influent line for eight seconds. Next, it begins to draw the sample in and discharges it out of the peristaltic to jar tubing. After collecting the sample, it again purges the influent line. When performing the calibration step do not allow the sample to be discharged to the 2.5-gallon sample jar. The operator must capture the volume using a calibrated sample container. During this calibration step the pump uses its current calibration curve to expel the sample amount indicated by the operator. For the sample size used in this project, this value is 400 mL.

Measure the calibration sample collected and enter the value into the Glacier controller iNDut panel. The iNDut panel then provides the option to re-perform the calibration step. If the sample collected was erroneous by greater than 5% the step must be repeated. If the sample size was of an appropriate size continue on.

- Select "No." Indicates the operator is done calibrating the sampler.

At this point sampling begins and will continue for the preset amount of time (i.e., 24-h in this case). During testing, calibration of the sample size should be performed every 8 hours. To perform this step press the ruler button on the controller iNDut panel and follow the calibration instructions as outlined above.

Once the sampling has begun it is recommended that the operator, in addition to the 8-h calibrations, periodically check the process to insure that the unit is operating correctly. Because of the high sensitivity of the measurements, disturbances in sample size or missed samples can nullify a 24-h data set.

Sampling of Blanks. For quality control purposes it is also required that blank DI samples are acquired with the Glacier sampling apparatus used during testing. With a properly cleaned sampling set-up initiate the Glacier sampling chest to draw in1 L of DI water. Store this 1-L sample in a dark amber glass bottle and set aside for shipment with the other active samples. This process is performed to identify any contamination that might be initiated by the process itself.

Packaging and Shipment of Sample

Immediately after the 24-h collection, it is necessary to package and ship the sample for lab analysis. The necessary materials for this task include four 1-L dark amber glass bottles, 1-cubitainer per sampling site, one 36-quart ice chest, one 50-quart ice chest, bubble wrap, duct tape, and one chain-of-custody (COC) form per ice chest. For lab analysis only 2 L of sample are required, however, for duplication purposes it is required that an additional 2 L are collected. Thus, 4 L of each 24 sample will be distributed to the glass containers. The remainder of the 2.5-gallon sample should then be transferred to the cubitainer for additional water quality analysis.

Use the bubble wrap to carefully package each dark amber glass bottle individually. Place the sample bottle onto the frozen bed of Blue Ice packs. Fill the chest as thoroughly as possible with the Blue Ice packs to ensure consistent sample temperature during shipment. Using a plastic mailing envelope fill out the appropriate COC form and place it into the ice chest. Use bubble wrap as needed to insure no glass samples are broken during delivery. Use duct tape to completely seal the ice chests shut.

For shipment of the plastic cubitainers samples no bubble wrap is required. Place the samples into the 50-quart ice chest and pack with frozen Blue Ice packs. Insert a completed COC form using a plastic mailing envelope and seal the ice chest with duct tape. The samples are now ready for shipment to the appropriate laboratories.

Clean up and Deassembly

After the 24-h sample sets have been collected and sent for shipment, the Glacier sampling chests are deassembled and cleaned. For doing this, break the tubing structure into the three original pieces (basin-to-peristaltic pump tubing, peristaltic pump tubing, and pump-to-sample jar tubing). Using the cleaning procedure outlined previously clean completely each section of sample tubing, the nozzle strainer, 2.5-gallon sample jar, and the interior of the Glacier sampling chest. For each piece of equipment, it is required to use both acetone and hexane to ensure sample integrity. After the sampling equipment is properly cleaned and packed away, the operator should contact the appropriate plant personnel to be advised regarding the method of disposing of the used cleaning chemicals. This requires filling out the facilities chemical disposal form for staff approval.

APPENDIX C

Operational Data Log Sheets For Plant Personnel

Plant Identification

Agency Name	
Plant Name	
Plant Address	
Name of Plant Operator	
Phone # of Plant Operator	
Name of Contact for Permission to Sample	
Phone # of Contact for Permission to Sample	
Name of Contact to Coordinate Site Visits for Sampling	
Phone # of Contact to Coordinate Site Visits for Sampling	

This space for any additional plant information:

Plant Flow Data

Flow Measuring Device	
Measurement Location	
Average Dry Flow (Design)	
Average Dry Flow (Currently)	
How can we obtain flow information for past year?	

Type of historical flow information available?

Plant Unit Processes (indicate all that apply)

Screening	
Comminution	
Grit Removal	
Off-line Flow Equilization	
In-line Flow Equilization	
Primary Settling	
Suspended Biological	
Attached Biological	
Denitrification	
Phosphorous Removal	
Secondary Settling	
Media Filtration	
Membrane Bioreactor	
Microfiltration	
Ultrafiltration	
Reverse Osmosis	
Electrodialysis	
Ozonation	
Chlorine Disinfection	
UV Disinfection	

Primary Process Operation During Sampling	
Shape of primary tank (circular or rectangular)	
Rectangular Dimensions (d x l x w)	
Circular Dimensions (d x diam x bottom slope)	
Design Detention Time	
Design Overflow Rate (Average Flow)	
Design Overflow Rate (Peak Flow)	

Secondary Process Operation During Sampling	
Influent Flow Rate (Q)	
Influent Temperature	
Aerobic Zone	
Return Sludge Flow Rate (Qw)	
Waste Sludge Flowrate (Q-Qw)	
Biological Reactor Volume (V)	
MLVSS in biological reactor (X)	
VSS in effluent and RAS (Xe and Xr)	
SRT[1] (calculated value)	
HRT[2] (calculated value)	
Anoxic Zone	
Return Sludge Flow Rate (Qw)	
Waste Sludge Flowrate (Q-Qw)	
Biological Reactor Volume (V)	
MLVSS in biological reactor (X)	
VSS in effluent and RAS (Xe and Xr)	
SRT[1] (calculated value)	
HRT[2] (calculated value)	
Organic Volumetric Loading Rate (Lorg)[3]	
Aeration DO Concentration	
Provide DO profile through the plant	
Aeration Device (circle one): Diffuser, Mechanical Aerator	
Fine or Coarse Bubble Aerator	
Formula Defintions	
[1] SRT = VX/((Q-Qw)Xe+QwXr	
[2] HRT = V/Q	
[3] Lorg = ((Q)(So))/((V)(10^3 g/kg)	

Tertiary Treatment Operation During Sampling	
Granular or Dynasand?	
Number and Size of Filter Units	
Continuous or Periodic Backwash	
Periodic Backwash Frequency	
Filtration Rate	
Chemical Addition Used?	
Chemical Type & Dose?	
Effluent Turbidity (Average)	

Membrane Treatment Operation During Sampling	
Low-Pressure Membrane Characteristics	
Manufacturer & Model Number (MF/UF)	
Pore Size of MF/UF	
Number of Modules & Elements	
Effective Surface Area per Module or Element	
Number of Skids or Cassettes	
Design Hydraulic Capacity of Plant	
Design Percent Feedwater Recovery	
Backwash Recovery Method	
Backwash Chemicals Used	
Maintenance Cleaning Chemicals Used	
How are membrane residuals disposed of	
Operating Characteristics During Sampling:	
Flux	
Transmembrane Pressure	
Applied Pressure	
Feedwater Recovery	
Daily Production	
High-Pressure Membrane Characteristics	
Manufacture & Model Number (NF/RO/Electrodialysis)	
RO Membrane Material	
Number of Modules per Vessel	
Number of Vessels per Stage	
Number of Stages	
Effective Surface Area per Module or Element	
Number of Skids or Cassettes	
Design Hydraulic Capacity of Plant	
Design Percent Feedwater Recovery	
Anti-scalant Used	
Chloramine Dose Used	
Maintenance Cleaning Chemicals Used	
How are membrane residuals disposed of	
Operating Characteristics During Sampling:	
Flux	
Permeate Pressure	
Concentrate Pressure	
Feedwater Recovery	
Daily Production	

Plant Identification	
Agency Name	
Plant Name	
Plant Address	
Name of Contact for Permission to Sample	
Phone # of Contact for Permission to Sample	
Name of Contact to Coordinate Site Visits for Sampling	
Phone # of Contact to Coordinate Site Visits for Sampling	
Name of MWH Person Sampling (print)	
Name of MWH Person Sampling (print)	
Plant Staff Assisting (print)	
Plant Staff Assisting (print)	
Date of Sampling	

Sample Collection Points and Collection Devices	
Collection Location One	
Date/Time Device Activated	
Date/Time Device De-activated	
Collection Device One (Make/model/owner)	
Cleaning Protocol	
Analytes Collected	
Chain of Custody Form Number	
24-hr Time or Flow Composite?	
If flow composite, how measured?	
Suction Tubing Material	
Pump or Pump Tubing Material	
Collection Location Two	
Date/Time Device Activated	
Date/Time Device De-activated	
Collection Device Two (Make/model/owner)	
Cleaning Protocol	
Analytes Collected	
Chain of Custody Form Number	
24-hr Time or Flow Composite?	
If flow composite, how measured?	
Suction Tubing Material	
Pump or Pump Tubing Material	
Collection Location Three	
Date/Time Device Activated	
Date/Time Device De-activated	
Collection Device Three (Make/model/owner)	
Cleaning Protocol	
Analytes Collected	
Chain of Custody Form Number	
24-hr Time or Flow Composite?	
If flow composite, how measured?	
Suction Tubing Material	
Pump or Pump Tubing Material	

Collection Location Four	
Date/Time Device Activated	
Date/Time Device De-activated	
Collection Device Four (Make/model/owner)	
Cleaning Protocol	
Analytes Collected	
Chain of Custody Form Number	
24-hr Time or Flow Composite?	
If flow composite, how measured?	
Suction Tubing Material	
Pump or Pump Tubing Material	
Record miscellaneous notes below:	

Water Quality Data from Plant on Date(s) Sampled *(Record "NA" if not available)*	
Raw BOD and date/time of sample collection	
2° Influent BOD and date/time of sample collection	
2° Effluent BOD and date/time of sample collection	
3° Effluent BOD and date/time of sample collection	
Influent TSS and date/time of sample collection	
Effluent TSS and date/time of sample collection	
Influent COD and date/time of sample collection	
Effluent COD and date/time of sample collection	
Influent VSS and date/time of sample collection	
Effluent VSS and date/time of sample collection	
Influent TKN and date/time of sample collection	
Effluent TKN and date/time of sample collection	
Influent NH3 and date/time of sample collection	
Effluent NH3 and date/time of sample collection	
Influent pH and date/time of sample collection	
Effluent pH and date/time of sample collection	
Temperature and date/time of measurement	

APPENDIX D

OPERATIONAL PARAMETERS OF SECONDARY TREATMENT AND WATER QUALITY ANALYSIS OF SECONDARY INFLUENT/EFFLUENT

Table D-1. Operational Parameters of Secondary Treatment from Participant Facilities.

Plant	Process	Sample ID	Sample Date	SRT, d		HRT, h		MLSS, mg/L		Tetum-Activated Sludge, % of influent	
				Sampling Day	Monthly Average	Sampling Day	Monthly Average	Sampling Day	Monthly Average	Sampling Day	Monthly Average
A	High-purity oxygen	A-1	8/16/2005	1.0	1.1	2.5	2.4	1520	1581	39	40
		A-2	3/2/2006	1.0	0.9	2.2	2.2	1550	1550	40	40
		A-3	3/6/2006	0.8	0.9	2.2	2.2	1550	1550	40	40
B	Conventional plug flow	B-1	3/9/2006	2.7	4	3.8	3.7	2220	2258	67	69
		B-2	3/14/2006	3.1	4	3.8	3.7	2018	2258	72	69
		B-3	3/15/2006	3.6	4	3.8	3.7	2075	2258	59	69
C	Conventional plug flow	C-1	9/27/2005	5.1	5.4	6.2	6.1	1220	1219	40	40
		C-2	11/16/2005	5.0	5.4	6.3	6.3	1104	1164	40	40
		C-3	1/25/2006	5.0	5.1	6.3	6.3	1176	1222	40	40
D	Conventional plug flow	D-1	9/22/2005	11.2	9.3	6.6	5.6	2375	2510	73	70
		D-2	11/21/2005	6.1	6.6	5.5	5.6	1903	2578	72	71
		D-3	2/16/2006	5.7	6.9	5.2	5.3	2757	2783	75	76
E[1]	Conventional plug flow	E-1	12/6/2005	12.3	16.3	28	30	1975	2108	80	80
		E-2	12/8/2005	12.9	16.3	30	30	1905	2108	80	80
F	Extended aeration	F-1	10/5/2005	27.7	27.8	24.7	25.2	4200	4306	70	70
		F-2	11/2/2005	31.3	28.4	24.4	24.2	4805	4410	70	70
		F-3	1/9/2006	29.5	29.2	21.6	21.1	4805	4845	70	70
MBR #01	MBR	M1-1	2/8/2006	8.5	14.5	6.6	6.6	11300	11600	NA[2]	NA
		M1-2	2/10/2006	12.3	14.5	6.6	6.6	11700	11600	NA	NA
		M1-3	2/11/2006	11.3	14.5	6.6	6.6	11700	11600	NA	NA
MBR #02[1]	MBR	M2-1	12/6/2005	15	15	5.1	5.1	13973	13872	NA	NA
		M2-2	12/8/2005	15	15	5.1	5.1	13665	13872	NA	NA

[1] Plant E and MBR #02 are located at the same site.
[2] NA = not applicable.

Table D-2. Secondary Influent Water Quality in Participant Facilities.

Sample ID	Sample Date	BOD, mg/L		TSS, mg/L		VSS, mg/L		TKN, mg/L	NH_3–N, mg/L	NO_2–N, mg/L	NO_3–N, mg/L
		Sampling Day	Monthly Average	Sampling Day	Monthly Average	Sampling Day	Monthly Average	Sampling Day	Sampling Day	Sampling Day	Sampling Day
A-1	8/16/2005	179	189	159	166	NA[1]	NA	55.6	35	<1	0.56
A-2	3/2/2006	179	188	154	142	NA	NA	46.2	35.9	<0.3	0.28
A-3	3/6/2006	186	188	151	142	NA	NA	48	39.3	<0.5	0.27
B-1	3/9/2006	128	126	57	65	NA	NA	NA	31.4[2]	NA	NA
B-2	3/14/2006	124	126	53	65	NA	NA	135	29.7	<1	0.62
B-3	3/15/2006	122	126	123	65	NA	NA	60.5	30.6	<1	<1.2
C-1	9/27/2005	113	120	96	89	75	72	39.7	37.2	<0.5	0.23
C-2	11/16/2005	140	139	81	80	64	64	40.6	30.1	<0.4	0.28
C-3	1/25/2006	156	147	85	84	79	79	NA	NA	NA	NA
D-1	9/22/2005	129	110	56	51	45	41	31.1	25.6	<2.5	1.3
D-2	11/21/2005	179	164	69	82	56	67	44.7	30.3	<2	<2.4
D-3	2/16/2006	195	176	105	87	84	67	NA	28[2]	<2	1.4
E-1	12/6/2005	251	316	210	212	198	182	34.6	31.2	<0.2	<0.24
E-2	12/8/2005	277	316	257	212	176	182	38.8	32.2	<0.13	0.082
F-1	10/5/2005	173	211	187	281	NA	NA	34.7	29	<0.05	0.045
F-2	11/2/2005	174	213	186	248	NA	NA	43.8	13.8	<0.4	0.26
F-3	1/9/2006	226	227	227	319	NA	NA	41	30.3	<0.4	<0.48
M1-1	2/8/2006	188	180	194	197	141	143	37.7	33.3	<1	0.67
M1-2	2/10/2006	172	180	194	197	142	143	NA	36.9	<1	0.6
M1-3	2/11/2006	172	180	194	197	142	143	38.5	31.4	<1	0.58
M2-1	12/6/2005	251	247	190	212	157	182	34.6	31.2	<0.2	<0.24
M2-2	12/8/2005	191	247	257	212	220	182	38.8	31.2	<0.13	0.082

[1] NA, not available.

[2] Reported by the plant.

Table D-3. Secondary Effluent Water Quality in Participant Facilities.

Sample ID	Sample Date	BOD, mg/L		TSS, mg/L		VSS, mg/L		TKN, mg/L	NH_3–N, mg/L	NO_2–N, mg/L	NO_3–N, mg/L
		Sampling Day	Monthly Average	Sampling Day	Monthly Average	Sampling Day	Monthly Average	Sampling Day	Sampling Day	Sampling Day	Sampling Day
A-1	8/16/2005	21	20	25	23.7	NA[1]	NA	40.2	NA	<1	0.83
A-2	3/2/2006	24	22	24	22.7	NA	NA	36.8	33.9	<0.3	0.31
A-3	3/6/2006	21	22	23	22.7	NA	NA	36.2	34.4	<0.5	0.31
B-1	3/9/2006	12.8	16.6	8.3	9.2	NA	NA	NA	18[2]	NA	NA
B-2	3/14/2006	17.6	16.6	6.9	9.2	NA	NA	14	12	0.43	1.4
B-3	3/15/2006	20	16.6	6.5	9.2	NA	NA	17.7	16.2	4	0.91
C-1	9/27/2005	7.5	9.3	6.2	7.3	5.4	6.1	5.5	4.2	5.3	10.1
C-2	11/16/2005	8.3	8.4	6.5	7.1	6.5	6.5	8.9	8.1	4.5	9
C-3	1/25/2006	12.5	12.8	11.8	11.5	10.3	10.1	5.8	5	2.2	8.7
D-1	9/22/2005	6	<4	6	5	NA	NA	2	0.23	<2.5	8.2
D-2	11/21/2005	6	<5	9	6	NA	NA	3	0.9	<2	6
D-3	2/16/2006	22	15	32	21	NA	NA	2.6	0.92	<2	9.4
E-1	12/6/2005	2.2	2.2	10.4	7.3	NA	NA	1.1	0.45	<0.25	12.5
E-2	12/8/2005	2.12	2.2	16.4	7.3	NA	NA	1	0.85	<0.25	14
F-1	10/5/2005	1.0	1.3	0.2	0.8	NA	NA	1.7	1.6	<0.5	0.33
F-2	11/2/2005	1.1	1.3	0.1	0.8	NA	NA	1.6	0.53	<0.4	0.45
F-3	1/9/2006	0	0.1	0	0.1	NA	NA	1.9	0.87	<0.4	0.58
M1-1	2/8/2006	<2	<2	NA	NA	NA	NA	0.83	0.4	<1	31
M1-2	2/10/2006	<2	<2	NA	NA	NA	NA	0.55	0.11	<1	32.5
M1-3	2/11/2006	<2	<2	NA	NA	NA	NA	0.56	0.2	<1	29.7
M2-1	12/6/2005	<4	<4	NA	NA	NA	NA	21.9	13.1	<0.2	0.11
M2-2	12/8/2005	<4	<4	NA	NA	NA	NA	30.6	27.7	<0.13	0.069

[1] NA, not available.

[2] Reported by the plant.

APPENDIX E

GENERAL QA/QC PROCEDURES

A. Quality Objectives for Data Measurement

A Quality Assurance and Quality Control (QA/QC) plan was developed to ensure the accuracy of data generated during the full-scale and pilot-scale sampling and analysis. This included results of samples for physical and chemical analysis and measurements of operational parameters. Water quality samples and measurements were collected from multiple plants and at frequencies sufficient to demonstrate representative plant operating conditions. Where standard methods exist for analyses, they were followed. QA/QC procedures were implemented to assure both precision and accuracy for all water quality analysis. Accordingly, MDLs were identified and matrix spikes, duplicates, and blanks were performed.

This QA/QC plan identified acceptance criteria and corrective action for all water quality parameters. Sample handling procedures included proper preservation and storage methods for all collected samples and the use of chain of custody (COC) forms. Samples were delivered and analyzed within the allotted holding time of each parameter. Data collection forms were developed to ensure that all required parameters are recorded at the time of sampling. Computerized data files were backed up on a regular basis. Standardized procedures for the collection, handling, and storage of data were developed to ensure the electronic data used in developing conclusions at the end of the project were the same as the data originally generated.

Criteria for the acceptance or rejection of data were based on following the quality objective procedures developed for sampling, analysis, and data management. Data generated under the conditions of quality objective procedures being followed were accepted. Data generated under conditions where quality objective procedures were not followed were considered on a data point-by-data point basis. The detailed QC criteria are presented in the following sections.

B. Quality Control Requirements

Data quality for specified sampling protocols was achieved by meeting the QC requirements of analytical methods and calibration and verification of instruments and equipment. Because of the distinct nature of the analyses conducted in the field (i.e., during sampling events) and those conducted in the lab, there was a need for separate QA/QC procedures for each case.

B.1 Field QA/QC procedures

<u>General</u>

Equipment calibration and verification procedures and frequencies are listed in the section named "Instrument/Equipment Testing." Equipment and instrument verification results that were greater than 10% different from expected and affected existing data were reported. Equipment and instruments were calibrated and reverified after verification results outside the 10% requirement. These considerations applied to pH meters, thermometers, sampling devices, and MBR operating controls. The project dedicated composite sampling devices were thoroughly tested before the sampling event. Proper functioning of all mobile parts was verified and Teflon tubing and couplings were used in the pump inlet and outlet to avoid contamination with phthalates (i.e., compounds commonly present in regular hoses). Medical-grade PVC tubing was used inside the peristaltic pump.

After transferring the sample volume needed for the PPCP analyses (2 L), aliquots for customary water quality analyses (BOD, TSS, VSS, TKN, NH_3–N, NO_2–N, NO_3–N) were transferred from the remainder of the 10-L composite sampler to sample bottles provided by the commercial laboratory. This process was conducted even if these data were available from the plant, in which case, it served as an additional data set. Customary water quality data were obtained for the secondary process influent and effluent samples and nitrate and nitrite parameters were substituted for TKN for the effluent sample. Samples for each analysis were collected in the appropriate sampling bottles, placed in a cooler with sufficient "blue ice" (to keep the temperature at a maximum of 4 °C), and sent to the predetermined lab for same day analysis.

Related to PPCP lab analysis

One blank and one duplicate sample were collected per plant sampling event. These samples were collected following the standard procedure (adapted from Snyder, 2005) presented in Appendix B.

Blanks The QA/QC procedure consisted of running laboratory-grade distilled water through one of the sampling devices during each sampling event, before any sample collection. This water was transferred to the PPCP amber bottles and constituted the blank sample for each sampling event.

Duplicates A duplicate split sample will be selected for one of the sampling locations at each sampling event. This location will be rotated during subsequent seasonal sampling of each facility.

B.2 Analytical Lab QA/QC procedures

This subsection describes all procedures used to ensure that all lab results from PPCP analysis are as meaningful and accurate as possible. The analytical methods for PPCP analysis were described in the Section 2.3.2. One blank sample and one duplicate sample were collected per plant sampling event.

Blanks

The extraction method was already tested using blank samples consisting of laboratory-grade distilled water. Several compounds were identified in the water blanks, namely phenylpropionate, benzophenone, triphenylphosphate, butylbenzyl phthalate, and bis(2-ethylhexyl) phthalate. All were below the IDLs with the exception of bis(2-ethylhexyl) phthalate showing concentrations of 4 to 5 μg/L (IDLs were determined for pure compounds in ethyl acetate in accordance with U.S. EPA's 1993 guidelines) and were previously presented in Table 2-5. The major source of the phthalic acid esters was determined to be the SPE disks. The method was modified to include a step were the SPE disks were washed with three aliquots of ethyl acetate before the conditioning of the disks. Although this reduced the level of contamination the concentrations of bis(2-ethylhexyl) phthalate remained high. The U.S. EPA guidance is that a compound cannot be reported if it is found in the field blank unless its concentration is 10 times greater than that of the blanks (Till, 2003). For example, bis(2-ethylhexyl) phthalate was found in the blank water samples at a concentration of 4.8 μg/L. Therefore, the reporting limit for bis(2-ethylhexyl) phthalate would be 48 μg/L.

Duplicates

For each batch of samples, one comparison of a sample and its respective duplicate was performed. Total area counts of both replicates must be within 20% of each other. If this criterion failed, all samples in the batch were reanalyzed.

Surrogates

Four surrogates were used, caffeine-$^{13}C_3$, 3, 3-dichlorobenzidine-d_3, 2, 4-dichlorophenol-d_3, and bis(2-ethylhexyl) phthalate-d_4 (Cambridge Isotope Laboratories, Andover, Mass.). This group of chemicals provided a range of polarities and acidities similar to the analytes. The surrogate recoveries must be greater than 70% and less than 130%. If the surrogates failed these criteria the sample was reanalyzed.

B.3 Instrument/equipment testing inspection and maintenance

Required maintenance of all instruments and equipment was performed in accordance with manufacturer's guidelines. Maintenance logs of routine inspections and corrective actions were implemented. Instruments were visually checked for proper operation on a daily basis when in use. When required, discrepancies and corrective actions were recorded in the project logbook.

B.4 Instrument/equipment calibration and frequency

Field equipment

The sampling device(s) were tested and calibrated before conducting each sampling event. Thermometers and pH meters were calibrated to manufacturer's specifications and at the manufacturer recommended frequency. All samples collected for routine water quality analysis were sent to a certified laboratory where the application of proper QA/QC procedures for all instrument/equipment calibrations was guaranteed in advance.

Laboratory Analytical Equipment

This section is presented in Section 2.3.2.

B.5 Inspection and acceptance of supplies and consumables

Project staff at each test site was responsible for the inspection and acceptance of supplies and consumables. Test site staff coordinated with the project engineer on the provision of the exact supplies and consumables that will be used during the project. The project engineer was notified of any discrepancy in the received materials and/or consumables, and was responsible for determining the appropriate action to take in these cases.

B.6 Data acquisition requirements

Lab data were submitted to the project engineer using the standard reporting procedures applied by the laboratories (i.e., Dynaflow, Jessup, Md., and Applied Physics and Chemistry Laboratory, Chino, Calif.). In addition, the laboratories were in charge of ensuring the accuracy of all PPCP data provided to the project engineer. These data were subject to thorough QA/QC procedures before being released to the project engineer. The laboratories will keep records of all submitted data for further verification as necessary.

B.7 QA/QC procedures

The objective of this task was to establish a viable structure for the recording and transmission of data collected during the project. Information collected from the log sheets was transferred to computer spreadsheets and stored for further use during data analysis. The computer files were subject to QA/QC reviews to ensure that their content exactly reflects the collected data from the log sheets.

When applicable, online pilot-testing data recorded by data acquisition systems, such as operational and water quality parameters, was downloaded by manual importation into files in spreadsheet form. The data were manipulated into a convenient framework to allow analysis of membrane equipment operation. At a minimum, backup of the computer databases was performed on a weekly basis. For those parameters not recorded by the data acquisition system, field-testing operators recorded data and calculations by hand in laboratory notebooks. Daily measurements were recorded on specially prepared data log sheets. The original notebooks and specially prepared data log sheets were stored on-site; and photocopies were forwarded to the project engineer at least once every two weeks during testing. This protocol not only facilitated referencing the original data, but also offered protection of the original record of results. Pilot operating logs included a description of the membrane equipment operation (description of test runs, description of any problems or issues, etc.) in addition to experimental calculations and other items.

In general, a run number was assigned to each sampling event, and tied to the information from each step of data entry and/or analysis. This system was consistently used during the entire study to facilitate data tracking for the required purposes.

APPENDIX F

METHOD DETECTION LIMIT STUDY

There was concern that certain compounds, present in the wastewater sample used for the MDL calculations, were present at background concentrations that were too high which would lead to calculation of optimistically low MDLs. We have elected to review the RPD obtained for the split samples collected during each sampling event. Approximately 10 times above the statistically calculated MDL, there is sufficient precision to adequately quantify within acceptable method precision, typically 20–30%. Near the MDL value, there is only a low probability of generating false positives or false negatives and accuracy of the reported value can still be off by 100%. Therefore, if the existing MDL levels have been appropriately set, the RPD obtained for detects in the region around 10 times the MDL should agree within 20–30%, whereas those obtained at one to two times the MDL should agree within 100%. RPD data obtained only at values significantly above ten times the MDL value are not useful in this assessment according to the Revised Assessment of Detection and Quantitation Approaches, EPA-821-B-04-005.

The split sample data available from nine sampling events, summarized in Table I-1, demonstrate that for the majority of the target analytes appropriate MDL values have been set. Three compounds have insufficient detected data at this time to enable this assessment to be completed (methyl-3-phenylpropionate, ethyl-3-phenylpropionate, and triphenylphosphate). Of the remaining compounds, only octylmethoxycinnamate indicates a potential problem with the established MDL.

Table F-1. Review of Split Sample Data from Nine Sampling Events to Evaluate MDL Values.

Compound	RPD (%)	Ratio of Split Mean to MDL	Meets Expected Criteria*
Methyl-3-phenylpropionate	6.1	34	Indeterminate – need more data
3-Phenylpropionate	0.1 0.93 0.31	12 41 30	Yes
Ethyl-3-phenylpropionate	NA	NA	Indeterminate – need more data
Chloroxylenol	33 3.2 17 13 1.2 16 2.7	10 10 3.0 2.0 11 17 42	Yes
Methylparaben	13 10 2.6	1.8 17 15	Yes
BHA	136 15	4 22	Yes
DEET	1.6 4.9 13	2.9 3.8 6.3	Yes
Benzophenone	36 20 22 7.4 5.7 20 16	3.3 1.2 2.5 1.9 2.4 8.6 9.9	Yes
TCEP	2.9 3.8 9.8	2.9 1.5 2.3	Yes
Octylphenol	8.7	12	Yes
Benzylsalicylate	26 17 7.3	5.8 22 31	Yes
Caffeine	35 0 1.81	5.2 34 63	Yes
Galaxolide	13 8.4 18 1.8	2.5 3.1 2.0 3.8	Yes
Musk ketone	7.4 7.9 4.2 4 19	2.0 2.2 2.1 3.7 2.0	Yes
Butylbenzyl phthalate	61 23 1.2 0.58	2.8 3.1 6.2 26	Yes

Table F-1 (continued). Review of Split Sample Data from Nine Sampling Events to Evaluate MDL Values.

Compound	RPD (%)	Ratio of Split Mean to MDL	Meets Expected Criteria*
oxybenzone	28 3.9 2.6 6.1 11 6.3 2.8	8.9 4.2 3.2 2.8 12 35 29	Yes
triclosan	46 54 17 13 2.5 18 8.4	3.0 1.4 3.9 4.2 14 16 24	Yes
octylmethoxycinnamate	80 15 12	11 28 88	No
triphenylphosphate	NA	NA	Indeterminate – Need more data
Bis(2-ethylhexyl) phthalate	53	1.1	Yes
ibuprofen	11.4	6.7	Yes

* Expected criteria met by demonstrating RPD values not exceeding 30% at a split mean to MDL ratio of approximately 10 and not exceeding RPD values of 100% at a split mean to MDL ratio of approximately 1.

APPENDIX G

Percent Removal of PPCPs Through Tertiary Filters

Table G-1. Percent Removal of PPCPs through Tertiary Filters.

PPCP [1]	Plant F Filter (% removal)	Plant D Filter (%removal)	Plant D RO (% Removal)
chloroxylenol	-27[2]	<-147[3]	>59
	0	>42[4]	
	ND[5]	ND	
Methylparaben	>97	ND	
	ND	ND	
	ND	ND	
DEET	26	-8	>48
	ND	ND	
	ND	ND	
Benzophenone	<-88	-12	41[6]
	ND	-23	>38
	ND	ND	
TCEP	12	ND	
	-17	-100	>65
	ND	ND	
Galaxolide	ND	8	>42
	ND	-37	>46
	-18	-51	>77
Musk ketone	12	-2	>43
	0	>55	
	ND	-20	>68
Oxybenzone	3	ND	
	0	>68	
	ND	ND	
Triclosan	67	ND	
	15	-27	>93
	ND	-74	>79
Butylbenzyl phthalate	-24	ND	
	>4	ND	
	ND	ND	

[1] This table only presents the compounds that were detected in the secondary effluent.
[2] Negative percent removal indicates that the filter effluent concentration is greater than the filter influent concentration.
[3] <, indicates that the removal is smaller than the percentage value presented because the influent concentration is below MDL.
[4] >, indicates that the removal is greater than the percentage value presented because the effluent concentration is below MDL.
[5] ND, not detected both in influent and effluent
[6] 0.050 μg/L was detected in the RO effluent, however, 0.048 μg/L was found in the sampling blank (MDL = 0.043 μg/L)

APPENDIX H

OCCURRENCE OF PPCPs IN INFLUENT TO PLANTS WITHOUT PRIMARY TREATMENT

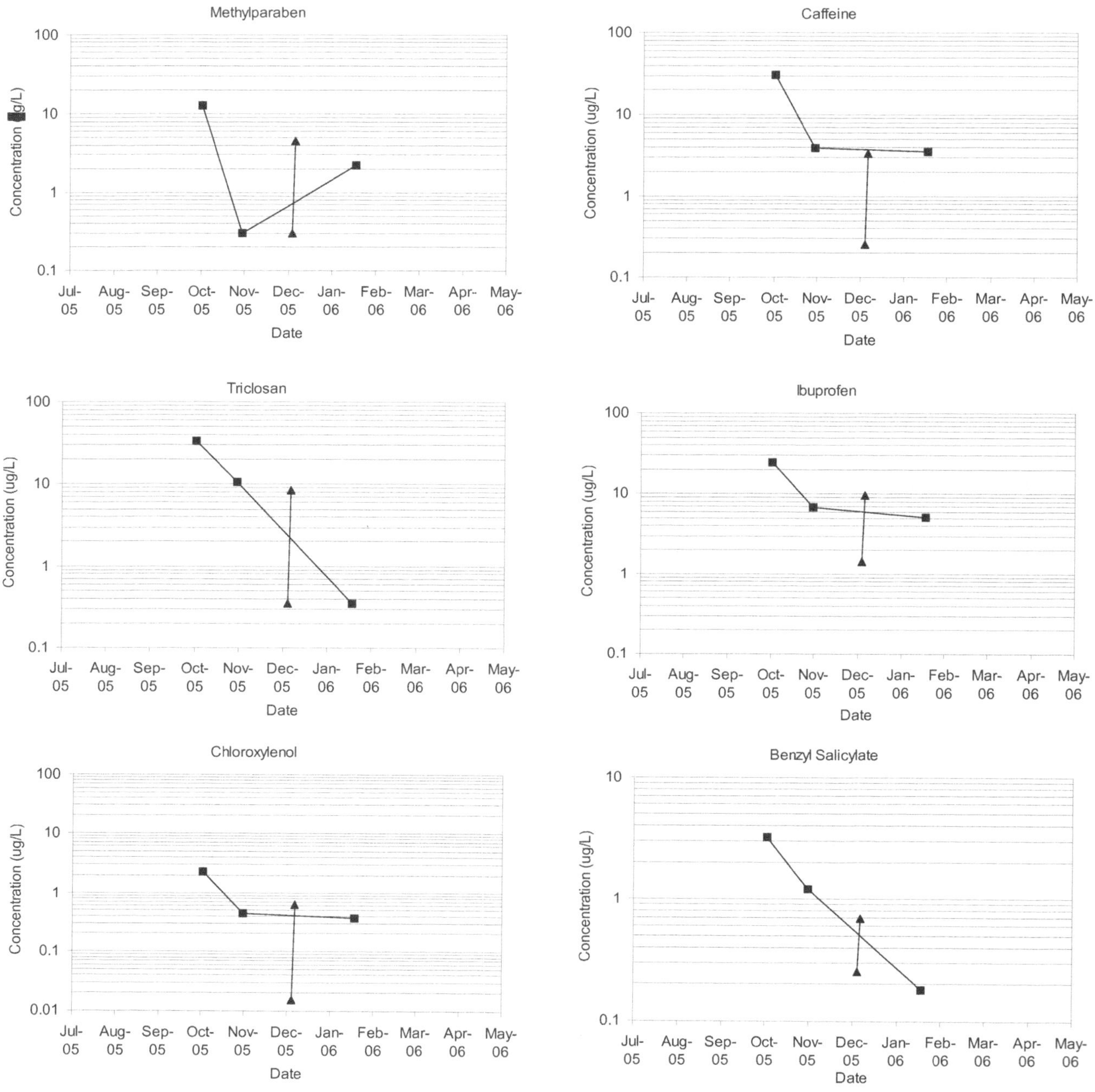

▲--- Plant E; ■ --- Plant F;
lines only to facilitate observation.

Figure H-1. Occurrence of PPCPs in Influent to Secondary Treatment in Plants without Primary Treatment.

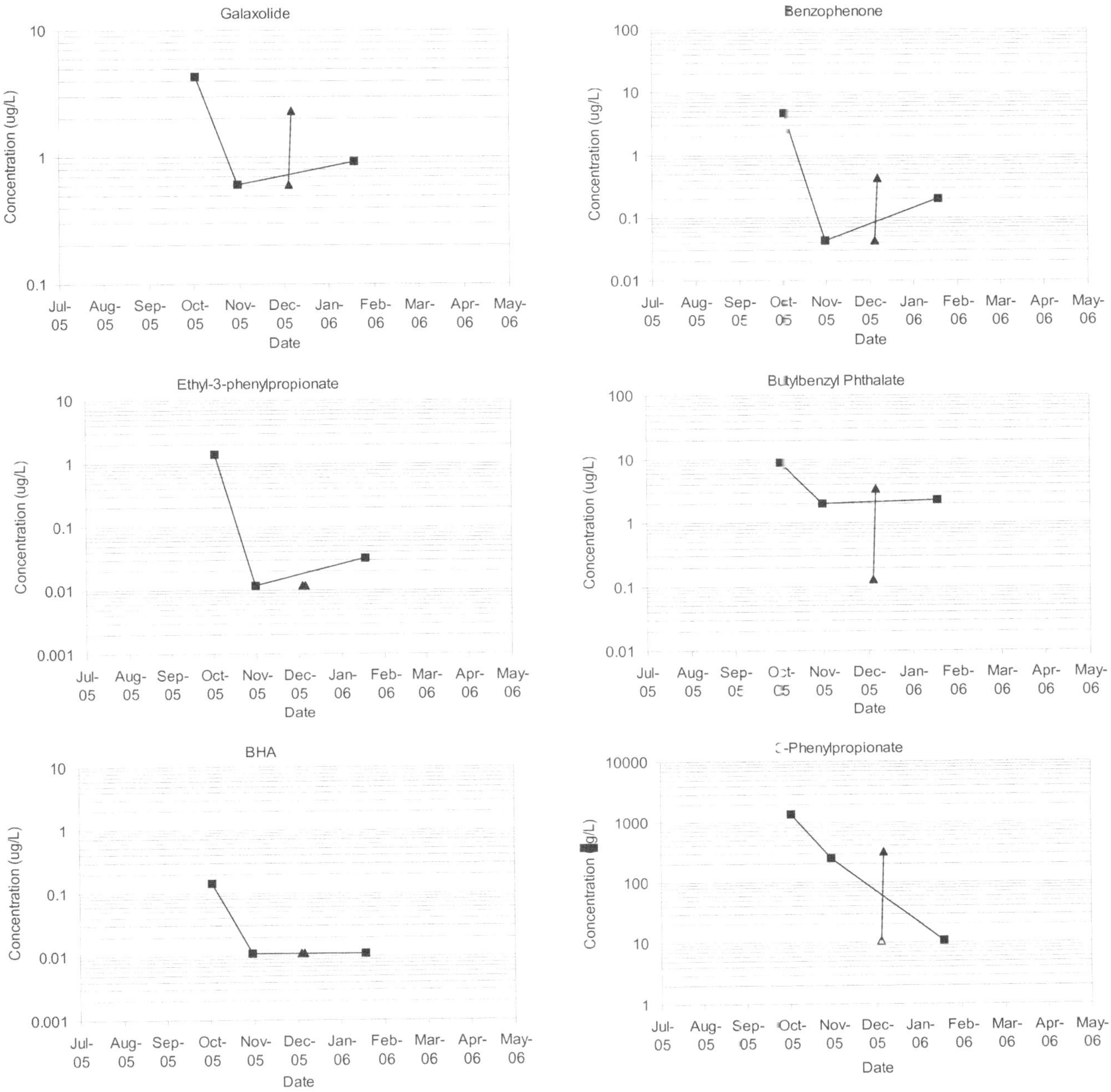

▲--- Plant E; ■ --- Plant F;
lines only to facilitate observation.

Figure H-1 (continued). Occurrence of PPCPs in Influent to Secondary Treatment in Plants without Primary Treatment.

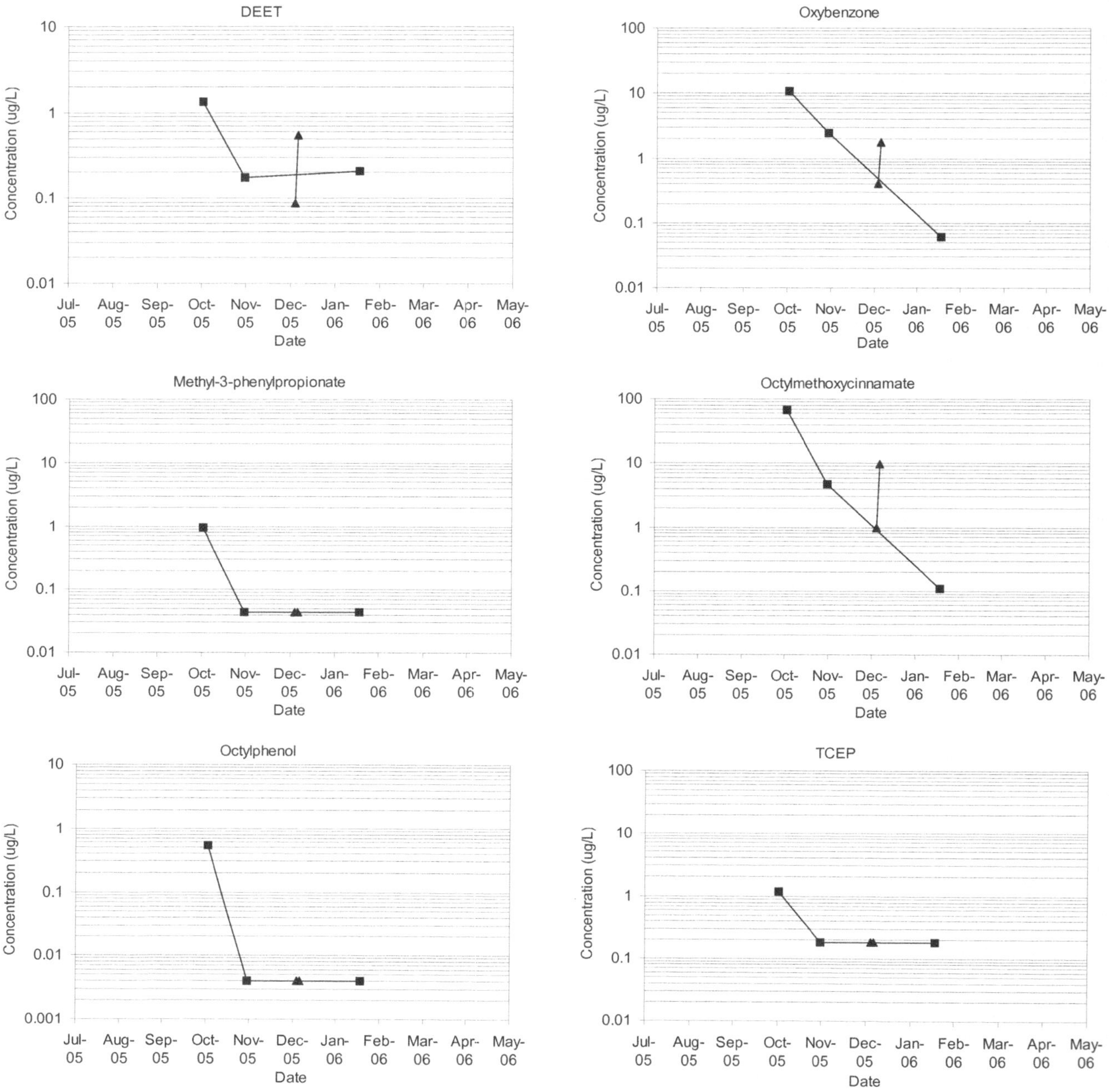

▲--- Plant E; ■ --- Plant F;
lines only to facilitate observation.

Figure H-1 (continued). Occurrence of PPCPs in Influent to Secondary Treatment in Plants without Primary Treatment.

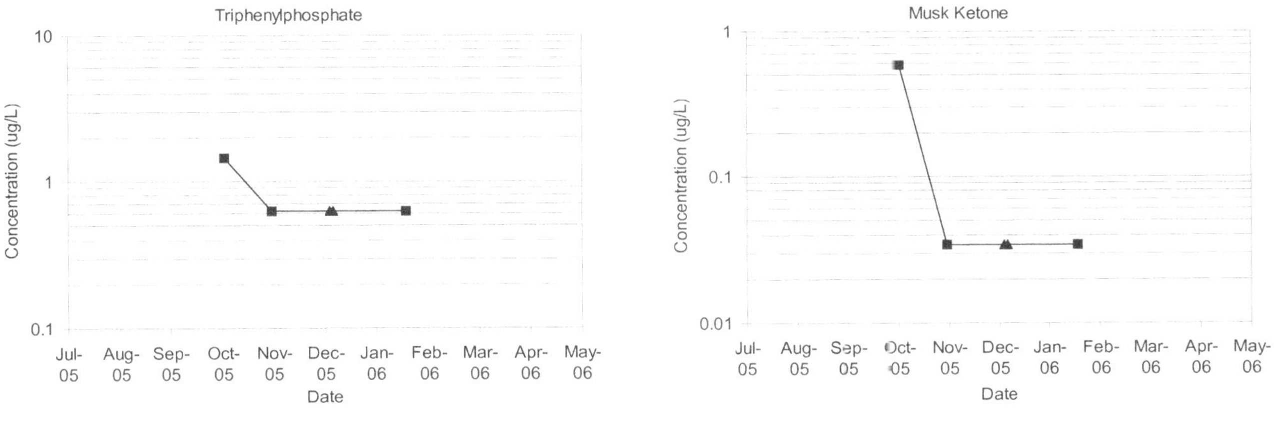

▲--- Plant E; ■ --- Plant F;
lines only to facilitate observation.

Figure H-1 (continued). Occurrence of PPCPs in Influent to Secondary Treatment in Plants without Primary Treatment.

APPENDIX I

OCCURRENCE OF PPCPS IN INFLUENT TO PLANTS WITH PRIMARY TREATMENT

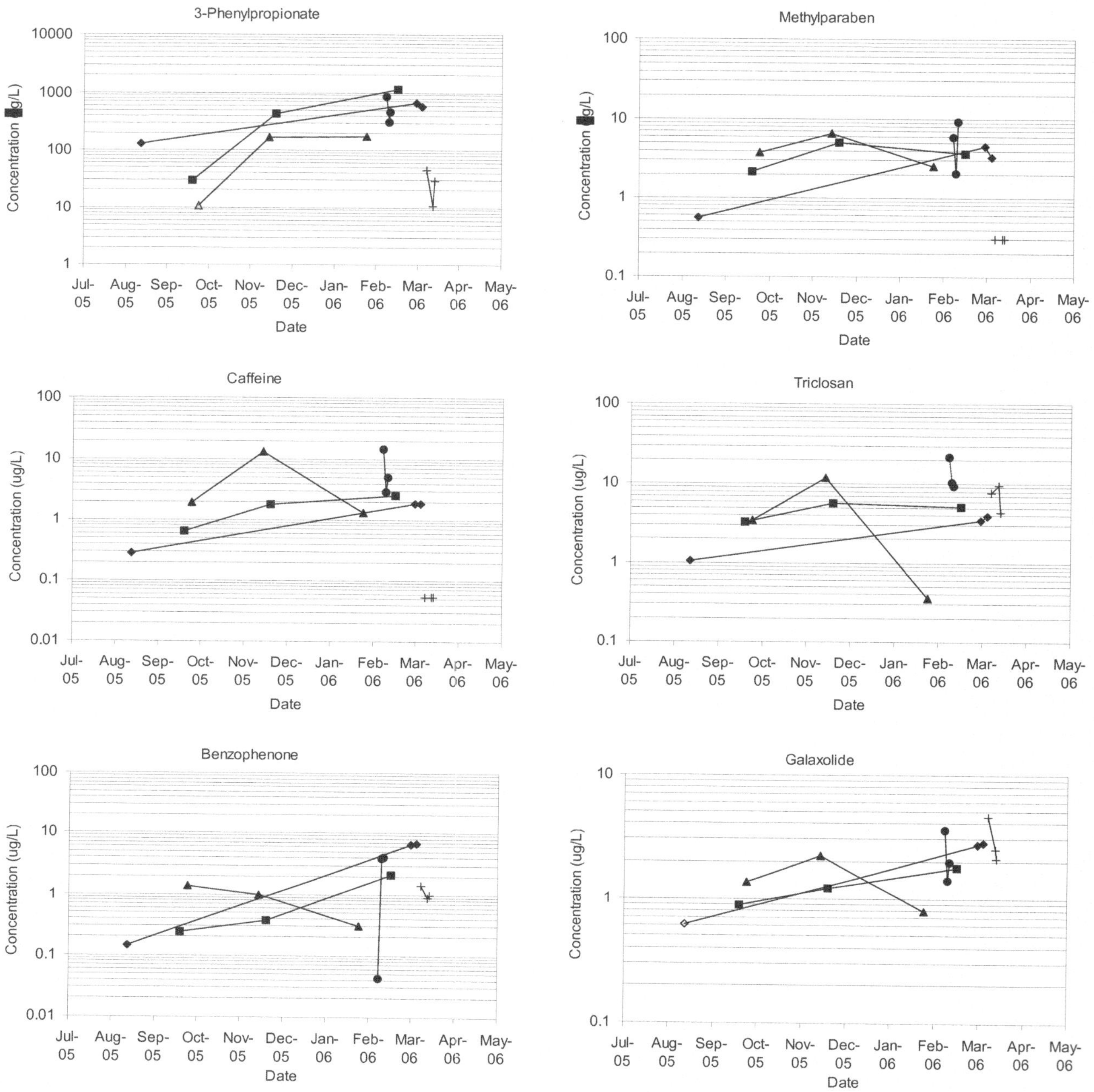

♦--- Plant A; + --- Plant B; ▲--- Plant C; ■ --- Plant D; ● --- MBR #01;
lines only to facilitate observation.

Figure I-1. Occurrence of PPCPs in Influent to Secondary Treatment in Plants with Primary Treatment.

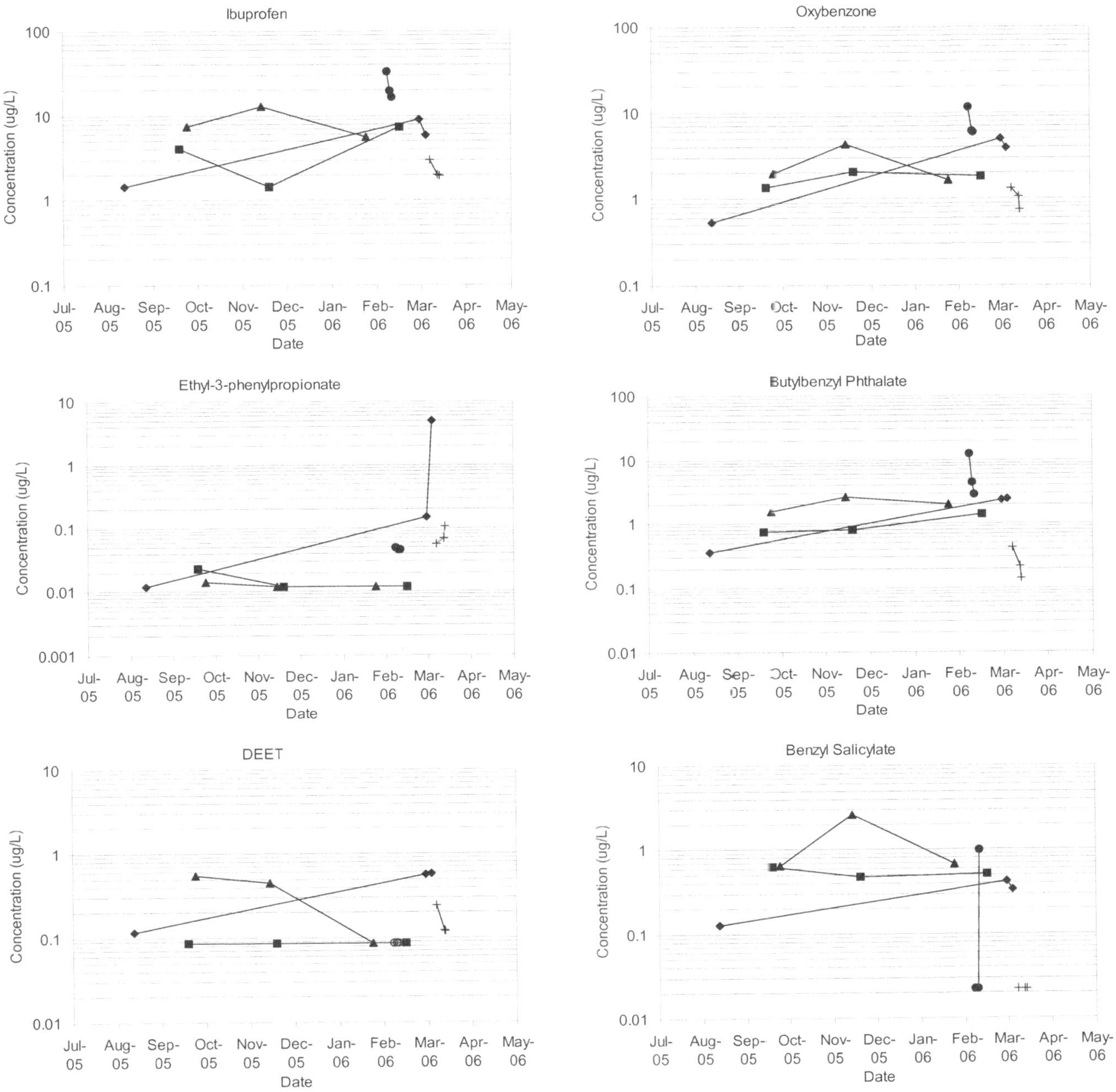

◆--- Plant A; + --- Plant B; ▲--- Plant C; ■ --- Plant D; ● --- MBR #01;
lines only to facilitate observation.

Figure I-1 (continued). Occurrence of PPCPs in Influent to Secondary Treatment in Plants with Primary Treatment.

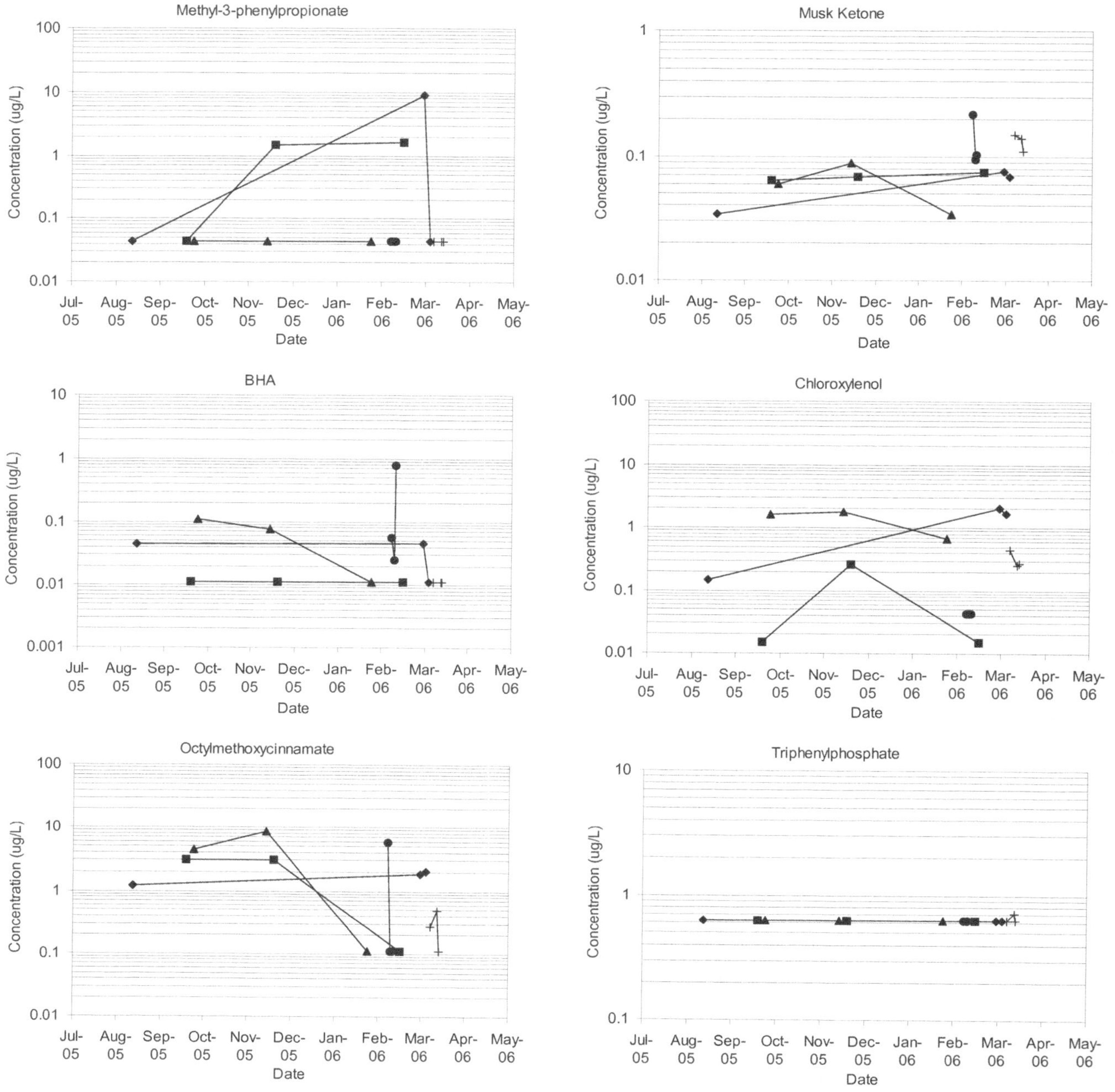

♦--- Plant A; + --- Plant B; ▲--- Plant C; ■ --- Plant D; ● --- MBR #01;
lines only to facilitate observation.

Figure I-1 (continued). Occurrence of PPCPs in Influent to Secondary Treatment in Plants with Primary Treatment.

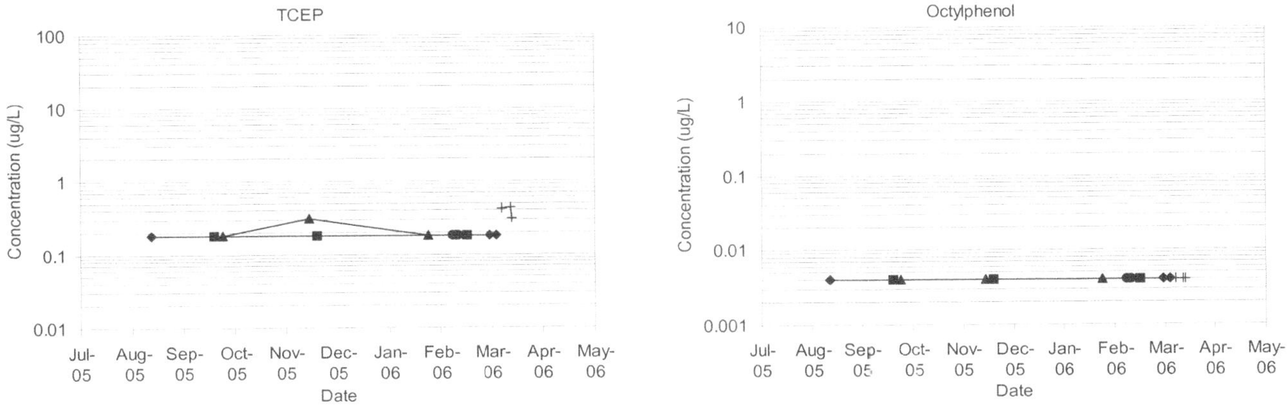

♦ --- Plant A; + --- Plant B; ▲ --- Plant C; ■ --- Plant D; ● --- MBR #01;
lines only to facilitate observation.

Figure I-1 (continued). Occurrence of PPCPs in Influent to Secondary Treatment in Plants with Primary Treatment.

References

Ahel, M.; Giger, W. Partitioning of Alkylphenols and Alkylphenol Polyethoxylates Between Qater and Organic-Solvents. *Chemosphere* **1993,** *26*(8), 1471–1478.

Andersen, H. Siegrist, H. Halling-Sorensen, B, Ternes, T. Fate of Estrogens in a Municipal Sewage Treatment Plant. *Environ. Sci. Tech.* **2003**, *37*(18), 4021–4026.

Balk, F.; Ford, R.A. Environmental Risk Assessment for the Polycyclic Musks AHTN and HHCB in the EU. I. Fate and Exposure Assessment. *Toxicol. Lett.* **1999,** *111*, 57–79.

Clara, M.; Kreuzinger, N.; Strenn, B.; Gans, O.; Droiss, H. The Solids Retention Time – A Suitable Design Parameter to Evaluate the Capacity of Wastewater Treatment Plants to Remove Micropollutants. *Water Res.* **2005**, *39*(1), 97–106.

Clara, M.; Strenn, B.; Auusserleitner, M.; Kreuzinger, N. Comparison of the Behavior of Selected Micropolllutants in a Membrane Bioreactor and a Conventional Wastewater Treatment Plant. *Water Sci. Tech.* **2004**, *50*(5), 29–36.

Daughton C.G.; Ternes T.A. Pharmaceutical and Personal Care Products in the Environment: Agents of Subtle Change? *Environ. Health Persp.* **1999**, *107*(6), 907–938.

Eriksson, E.; Auffart, K.; Eilersen, A-M, Henze, M.; Ledin, A. Household Chemicals and Personal Care Products as Sources for Xenobiotic Organic Compounds in Grey Wastewater. *Water SA*, **2003**, *29*(2), 135–146.

Everest, W.R.; Alexander, K.L.; Deshmukh, S.S.; Patel, M.V.; Daugherty, J.L.; Herberg, J.D. Emerging Contaminant Removal Using Reverse Osmosis for Indirect Potable Use. In *Proceedings of the IDA World Congress on Desalination and Water Reuse*, Paradise Island, Bahamas, 2003.International Desalination Association: New York, 2003.

Fromme, H.; Kuchler, T.; Otto, T.; Pilz, K.; Muller, J.; Wenzel, A. Occurrence of Phthalates and Bisphenol A and F in the Environment. *Water Res.* **2002**, *36*(6), 1429–1438.

Hansch, C.; Leo, A.; Hoekman, D. *Exploring QSAR;* American Chemical Society: Washington, DC, 1995.

Halling-Sørenson, B.; Nors Nielsen, S.; Lanzky, P.F.; Ingerslev, F.; Holten-Lützhøft, H.C.; Jergensen, S.E. Occurrence, Fate and Effects of Pharmaceutical Substances in the Environment – A Review. *Chemosphere* **1998**, *36*(2), 357–393.

The HERA Project. Human and Environmental Risk Assessment on Ingredients of Household Cleaning Products. Polycyclic Musks AHTN (CAS 1506-02-1) and HHCB (1222-05-05) – Report Version 2.0, HERA: Brussels, Belgium, 2004.

International Flavors & Fragrance. Robust Test Summaries for 1,3,4,6,7,8-Hexahydro-4,6,6,7,8,8-hexamethylcyclopenta-g-2-benzopyran (HHCB) CAS#1222-05-05; Report, IFF: New York, 2003.

Irwin, R. Source of Chemicals that Feminize Lake Mead Fish Discovered. In *National Park Service—Year in Review*, 1998, Selleck, J., Ed.; U.S. Dept. of the Interior: Lakewood, Colo., 1998. Available at http://www2.nature.NDs.gov/YearInReview/yir98/chapter04/chapter04pg5.html.

Jonsson, S.; Boren, H. Analysis of Mono- and Diesters of o-phthalic Acid by Solid Phase Extraction with Polystyrene-Divinylbenzene-based Polymers. *J. Chrom. A* **2002**, *963*, 393–400.

Joss, A.; Andersen, H.; Ternes, T.; Richie, P.; Siegrist, H. Removal of Estrogens in Municipal Wastewater Treatment under Aerobic and Anaerobic Conditions: Consequences for Plant Optimization. *Environ. Sci. Technol.* **2004**, 38 (11), 3047–3055.

Joss, A.; Keller, E.; Alder, A.C.; Gobel, A.; McArdell, C.S.; Ternes, T.; Siegrist, H. Removal of Pharmaceuticals and Fragrances in Biological Wastewater Treatment. *Water Res.* **2005**, 39: 3139–3152.

Kanda, R.; Griffin, P.; James, H.A.; Fothergill, J. Pharmaceutical and Personal Care Products in Sewage Treatment Works. *J. Environ. Monit.* **2003**, *5*, 823–830.

Kolpin, D.W.; Furlong, E.T.; Meyer, M.T.; Thurman, E.M.; Zaugg, S.D.; Barber, L.B.; Buxton, H.T. Pharmaceuticals, Hormones, and Other Organic Wastewater Contaminants in U.S. Streams, 1999–2000: A National Reconnaissance. *Environ. Sci. Technol.* **2002**, *36*(6), 1202–1211.

Kreuzinger, N.; Clara, M.; Droiss, H. Relevance of the Sludge Retention Time (SRT) as Design Criteria for Wastewater Treatment Plants for the Removal of Endocrine Disruptors and Pharmaceuticals from Wastewater. *Water Sci. Technol.* **2004**, *50*(5), 149–156.

Loraine, G.A. Effects of Seasonal Variations in PPCP Concentrations on Wastewater Treatment. Presented at American Chemical Society, 228th National Meeting, Philadelphia, Pa., August 22–26, 2004.

Loraine, G.A.; Pettigrove, M. Occurrence of pharmaceuticals and xenoestrogens in drinking water and wastewater in San Diego, Calif. Presented at CEA–Crest, 3rd Annual Environmental Science Conference, Pasadena Calif., May 3–4, 2002.

Ollers, S.; Singer, H, Fassler, P.; Muller, S. Simultaneous Quantification of Neutral and Acidic Pharmaceuticals and Pesticides at the Low-ng/l Level in Surface and Wastewater. *J. Chrom. A.* **2001**, *911*, 225–234.

Oros, D. Monitoring New Trace Organic Contaminants in 2002. *RMP Technical Report: Annual Monitoring Results*; San Francisco Estuary Institute: Oakland, Calif., 2002.

Osemwengie, L.I.; Steinberg, S. On-site Solid-Phase Extraction and Laboratory Analysis of Ultra-Trace Synthetic Musks in Municipal Sewage Effluent using Gas Chromatography-Mass Spectrometry in the Full-Scan Mode. *J. Chrom. A.* **2001**, *932*, 107–118.

Sanchez-Prado, L.; Lourido, M.; Lores, M.; Llompart, M.; Garcia-Jares, C.; Cela, R. Study of the Photoinduced Degradation of Polycyclic Musk Compounds by Solid-Phase Microextraction and Gas Chromatography/Mass Spectrometry. *Rapid Commun. Mass Spect.* **2004,** *18*(11), 1186–1192.

Sedlak, D.L.; Gray, J.; Pinkston, K.E. Understanding Microcontaminants in Recycled Water. *Environ. Sci. Technol.* **2000**, *34*(23), 508A–515A.

Snyder, E.M.; Snyder, S.A.; Kelly, K.L.; Gross, T.S.; Villaneuve, D.L.; Fitzgerald, S.G.; Villalobos, S.A.; Giesy, J.P. Reproductive Responses of Common Carp (*Cyprinus carpio*) Exposed in Cages to Influent of the Las Vegas Wash in Lake Mead, Nevada, from Late Winter to Early Spring. *Environ. Sci. Technol.* **2004**, *38*, 6385–6395.

Snyder, S. Personal communication, 2005.

Snyder, S.A.; Westerhoff, P.; Yoon, Y.; Sedlak, D.L. Pharmaceuticals, Personal Care Products and Endocrine Disruptors in Water: Implications for the Water Industry. *Environ. Eng. Sci.* **2003a**, *20*(5), 449–456.

Snyder, S.A.; Yoon, Y.; Westerhoff, P.; Vanderford, B.; Pearson, R.; Rexing, D. Evaluation of Conventional and Advanced Drinking Water Treatment Processes to Remove Endocrine Disruptors and Pharmaceutically Active Compounds: Bench-Scale Results. In *Proceedings of the 3rd International Conference on Pharmaceuticals and Endocrine Disrupting Compounds in Water;* The National Ground Water Association, Minneapolis, Minn, 2003b.

Snyder, S.; Kelly, K.; Grange, A.; Sovocool, G. W.; Snyder, E.; Giesy, J. Pharmaceuticals and Personal Care Products in the Waters of Lake Mead, Nevada. In *Pharmaceuticals and Personal Care Products in the Environment; Scientific and Regulatory Issues*, Daughton, C.; Jones-Lepp, T. Eds.; American Chemical Society: Washington, D.C., 2001**;** Chap. 7.

Tas, J.W.; Balk, F.; Ford, R.A.; van de Plassche, E.J. Environmental Risk Assessment of Musk Ketone and Musk Xylene in the Netherlands in Accordance with the EU-TGD. *Chemosphere* **1997**, *35*(12), 2973–3002.

Tchobanoglous, G.; Burton, F.L.; Stensel, H.D. *Wastewater Engineering: Treatment, Disposal and Reuse*; Fourth Ed.; McGraw-Hill Book Company: New York, 2003**.**

Ternes, T.A.; Stumpf, M.; Mueller, J.; Haberer, K.; Wilken, R.-D.; Servos, M. Behavior and Occurrence of Estrogens in Municipal Sewage Treatment Plants – I. Investigations in Germany, Canada and Brazil. *Sci. Tot. Environ.* **1999**, *225*, 81–99.

Till, A. Comment on Pharmaceuticals, Hormones, and Other Organic Wastewater Contaminants in U.S. Streams, 1999–2000: A National Reconnaissance. *Environ. Sci. Technol.* **2003**, *37*, 1052–1053.

Trussell, R.S.; Merlo, R.P.; Hermanowicz, S.W.; Jenkins, D. The Effect of High Mixed Liquor Suspended Solids Concentration, Mix Liquor Properties, and Coarse Bubble Aeration Flow Rate on Membrane Permeability. In *Proceedings of 2005 Water Environment Federation Technical Exhibition and Conference;* Washington, D.C., October 29–November 2, 2005; Water Environment Federation: Alexandria, Va., 2005.

U.S. Environmental Protection Agency (U.S. EPA).Guidelines Establishing Test Procedures for the Analysis of Pollutants: Procedures for Detection and Quantitation, Appendix B,Definition and Procedure. For the Determination of the Method Detection Limit; *Code of Federal Regulations,* Part 136, Title 40, 1993.

Wilson, B.A.; Smith, V.H.; Denoyelles, Jr., F.; Larive, C.K. Effects of Three Pharmaceutical and Personal Care Products on Natural Freshwater Algal Assemblages. *Environ. Sci. Technol.* **2003**, *37*, 1713–1719.

WASTEWATER UTILITY

Alabama
Montgomery Water Works & Sanitary Sewer Board

Alaska
Anchorage Water & Wastewater Utility

Arizona
Glendale, City of, Utilities Department
Mesa, City of
Peoria, City of
Phoenix Water Services Department
Pima County Wastewater Management
Safford, City of

Arkansas
Little Rock Wastewater Utility

California
Calaveras County Water District
Central Contra Costa Sanitary District
Corona, City of
Crestline Sanitation District
Delta Diablo Sanitation District
Dublin San Ramon Services District
East Bay Dischargers Authority
East Bay Municipal Utility District
Eastern Municipal Water District
El Dorado Irrigation District
Fairfield-Suisun Sewer District
Fresno Department of Public Utilities
Inland Empire Utilities Agency
Irvine Ranch Water District
Las Virgenes Municipal Water District
Livermore, City of
Los Angeles, City of
Los Angeles County, Sanitation Districts of
Napa Sanitation District
Orange County Sanitation District
Palo Alto, City of
Riverside, City of
Sacramento Regional County Sanitation District
San Diego Metropolitan Wastewater Department, City of
San Francisco, City & County of
San Jose, City of
Santa Barbara, City of
Santa Cruz, City of
Santa Rosa, City of
South Bayside System Authority
South Coast Water District
South Orange County Wastewater Authority
Stege Sanitary District
Sunnyvale, City of
Union Sanitary District
West Valley Sanitation District

Colorado
Aurora, City of
Boulder, City of
Greeley, City of
Littleton/Englewood Water Pollution Control Plant
Metro Wastewater Reclamation District, Denver

Connecticut
Greater New Haven WPCA

District of Columbia
District of Columbia Water & Sewer Authority

Florida
Broward, County of
Fort Lauderdale, City of
Miami-Dade Water & Sewer Authority
Orange County Utilities Department
Reedy Creek Improvement District
Seminole County Environmental Services
St. Petersburg, City of
Tallahassee, City of
Tampa, City of
Toho Water Authority
West Palm Beach, City of

Georgia
Atlanta Department of Watershed Management
Augusta, City of
Clayton County Water Authority
Cobb County Water System
Columbus Water Works
Fulton County
Gwinnett County Department of Public Utilities
Savannah, City of

Hawaii
Honolulu, City & County of

Idaho
Boise, City of

Illinois
American Bottoms Wastewater Treatment Plant
Greater Peoria Sanitary District
Kankakee River Metropolitan Agency
Metropolitan Water Reclamation District of Greater Chicago
Wheaton Sanitary District

Iowa
Ames, City of
Cedar Rapids Wastewater Facility
Des Moines, City of
Iowa City

Kansas
Johnson County Unified Wastewater Districts
Unified Government of Wyandotte County/ Kansas City, City of

Kentucky
Louisville & Jefferson County Metropolitan Sewer District
Sanitation District No. 1

Louisiana
Sewerage & Water Board of New Orleans

Maine
Bangor, City of
Portland Water District

Maryland
Anne Arundel County Bureau of Utility Operations
Howard County Department of Public Works
Washington Suburban Sanitary Commission

Massachusetts
Boston Water & Sewer Commission
Upper Blackstone Water Pollution Abatement District

Michigan
Ann Arbor, City of
Detroit, City of
Holland Board of Public Works
Saginaw, City of
Wayne County Department of Environment
Wyoming, City of

Minnesota
Rochester, City of
Western Lake Superior Sanitary District

Missouri
Independence, City of
Kansas City Missouri Water Services Department
Little Blue Valley Sewer District
Metropolitan St. Louis Sewer District

Nebraska
Lincoln Wastewater System

Nevada
Henderson, City of
Reno, City of

New Jersey
Bergen County Utilities Authority
Ocean, County of
Passaic Valley Sewerage Commissioners

New York
New York City Department of Environmental Protection

North Carolina
Charlotte/Mecklenburg Utilities
Durham, City of
Metropolitan Sewerage District of Buncombe County
Orange Water & Sewer Authority

Ohio
Akron, City of
Butler County Department of Environmental Services
Columbus, City of
Metropolitan Sewer District of Greater Cincinnati
Northeast Ohio Regional Sewer District
Summit, County of

Oklahoma
Oklahoma City Water & Wastewater Utility Department
Tulsa, City of

Oregon
Clean Water Services
Eugene, City of
Gresham, City of
Portland, City of
Water Environment Services

Pennsylvania
Philadelphia, City of
University Area Joint Authority

South Carolina
Charleston Water System
Mount Pleasant Waterworks & Sewer Commission
Spartanburg Water

Tennessee
Cleveland, City of
Knoxville Utilities Board
Murfreesboro Water & Sewer Department
Nashville Metro Water Services

Texas
Austin, City of
Dallas Water Utilities
Denton, City of
El Paso Water Utilities
Fort Worth, City of
Houston, City of
San Antonio Water System
Trinity River Authority

Utah
Salt Lake City Corporation

Virginia
Alexandria Sanitation Authority
Arlington, County of

Fairfax County
Hampton Roads Sanitation District
Henrico, County of
Hopewell Regional Wastewater Treatment Facility
Loudoun County Sanitation Authority
Lynchburg Regional WWTP
Prince William County Service Authority
Richmond, City of
Rivanna Water & Sewer Authority

Washington
Everett, City of
King County Department of Natural Resources
Seattle Public Utilities
Sunnyside, Port of
Yakima, City of

Wisconsin
Green Bay Metro Sewerage District
Kenosha Water Utility
Madison Metropolitan Sewerage District
Milwaukee Metropolitan Sewerage District
Racine, City of
Sheboygan Regional Wastewater Treatment
Wausau Water Works

Australia
South Australian Water Corporation
Sydney Water Corporation
Water Corporation of Western Australia

Canada
Regina, City of, Saskatchewan
Toronto, City of, Ontario
Winnipeg, City of, Manitoba

New Zealand
Watercare Services Limited

United Kingdom
Yorkshire Water Services Limited

STORMWATER UTILITY

California
Fresno Metropolitan Flood Control District
Los Angeles, City of, Department of Public Works
Monterey, City of
Sacramento, County of
San Francisco, City & County of
Santa Rosa, City of
Sunnyvale, City of

Colorado
Aurora, City of
Boulder, City of

Georgia
Griffin, City of

Iowa
Cedar Rapids Wastewater Facility
Des Moines, City of

Kansas
Overland Park, City of

Kentucky
Louisville & Jefferson County Metropolitan Sewer District

Maine
Portland Water District

North Carolina
Charlotte, City of, Stormwater Services

Pennsylvania
Philadelphia, City of

Tennessee
Chattanooga Stormwater Management

Texas
Harris County Flood Control District, Texas

Washington
Bellevue Utilities Department
Seattle Public Utilities

STATE

Arkansas Department of Environmental Quality
Connecticut Department of Environmental Protection
Kansas Department of Health & Environment
Kentucky Department of Environmental Protection
Ohio River Valley Sanitation Commission
Urban Drainage & Flood Control District, CO

CORPORATE

ADS Environmental Services
Advanced Data Mining International
Alan Plummer & Associates
Alden Research Laboratory Inc.
Alpine Technology Inc.
Aqua-Aerobic Systems Inc.
Aquateam–Norwegian Water Technology Centre A/S
ARCADIS
Associated Engineering
Black & Veatch
Boyle Engineering Corporation
Brown & Caldwell
Burgess & Niple, Ltd.
Burns & McDonnell
CABE Associates Inc.
The Cadmus Group
Camp Dresser & McKee Inc.
Carollo Engineers Inc.
Carpenter Environmental Associates Inc.
CDS Technologies Inc.
CET Engineering Services
Chemtrac Systems Inc.
CH2M HILL
Conestoga-Rovers & Associates
CONTECH Stormwater Solutions
D&B/Guarino Engineers, LLC
Damon S. Williams Associates, LLC
Earth Tech Inc.
Ecovation
EMA Inc.
Environ/The ADVENT Group, Inc.
Fay, Spofford, & Thorndike Inc.
Freese & Nichols Inc.
ftn Associates Inc.
Gannett Fleming Inc.
Garden & Associates, Ltd.
Geosyntec Consultants
GHD
Golder Associates Ltd.
Greeley and Hansen LLC
Hazen & Sawyer, P.C.
HDR Engineering Inc.
HNTB Corporation
Hydromantis Inc.
HydroQual Inc.
Infilco Degremont Inc.
Jacobson Helgoth Consultants Inc.
Jason Consultants LLC Inc.
Jordan, Jones, & Goulding Inc.
KCI Technologies Inc.
Kelly & Weaver, P.C.
Kennedy/Jenks Consultants
KMK Consultants
Komline Sanderson Engineering Corporation
Limno-Tech Inc.
Lombardo Associates Inc.
The Low Impact Development Center Inc.
Malcolm Pirnie Inc.
Material Matters
McKim & Creed
Metcalf & Eddy Inc.
Monteco Corporation
MPR Engineering Corporation, Inc.
MWH
NewFields Water Resources, LLC
O'Brien & Gere Engineers Inc.
Odor & Corrosion Technology Consultants Inc.
Oscar Larson & Associates
PA Government Services Inc.
Parametrix Inc.
Parsons
Post, Buckley, Schuh & Jernigan
Praxair, Inc.
RMC Water & Environment
Ross & Associates Ltd.
Rothberg, Tamburini & Windsor, Inc.
SAIC
Savin Engineers
Siemens Water Technologies
Stantec Consulting Inc.
Stearns & Wheler, LLC
Stone Environmental Inc.
Stratus Consulting Inc.
Synagro Technologies Inc.
Tetra Tech Inc.
Trojan Technologies Inc.
Trussell Technologies, Inc.
URS Corporation
Wade-Trim Inc.
Westin Engineering Inc.
Weston Solutions Inc.
Woodard & Curran
Zenon Environmental Inc.
Zoeller Pump Company

INDUSTRY

American Electric Power
American Water
ChevronTexaco Energy Research & Technology Company
The Coca-Cola Company
Dow Chemical Company
DuPont Company
Eastman Chemical Company
Eli Lilly & Company
Merck & Company Inc.
Premier Chemicals LLC
Procter & Gamble Company
Thames Water Utilities
Severn Trent Services Inc.
Suez Environnment
United Water Services LLC

Note: List as of 2/1/07